W0112734

Editors
Reinhard Bott and Thomas Langeloh

Solid/Liquid Separation Lexicon

Solid/Liquid Separation Lexicon

Editors
Reinhard Bott and Thomas Langeloh

Scientific Advisor
Harald Anlauf

Authors
Harald Anlauf
Reinhard Bott
Thomas Langeloh
Bernhard Hoffner
Klaus Julkowski
Franz Meck

Reinhard Bott
Thomas Langeloh
BOKELA Ingenieurgesellschaft für Mechanische Verfahrenstechnik mbH
Gottesauer Straße 28
D-76131 Karlsruhe

Library of Congress Card No. applied for.

A catalogue record for this book is available from the British Library.

Die Deutsche Bibliothek – Cataloguing-in-Publication Data
A catalogue record for this book is available from Die Deutsche Bibliothek

ISBN 3-527-30522-X

Printed on acid-free paper.

Composition: Stefanie Groß, Steinweiler
Printing: Strauss Offsetdruck GmbH, Mörlenbach
Bookbinding: Wilh. Osswald + Co. KG, Neustadt

Printed in the Federal Republic of Germany.

Editors

Reinhard Bott and Thomas Langeloh
BOKELA Ingenieurgesellschaft für Mechanische Verfahrenstechnik mbH
Gottesauer Straße 28
D-76131 Karlsruhe
Germany
Tel: +49 721 96 456-0
Fax +49 721 96 456-10
E-Mail: bokela@bokela.com
www.bokela.com

Scientific Advisor

Dr. Harald Anlauf
Institut für Mechanische Verfahrenstechnik und Mechanik (MVM)
Kaiserstraße 12
Universität Karlsruhe (TH)
D-76128 Karlsruhe
Germany

Authors

Dr. Reinhard Bott
Dr. Thomas Langeloh
Franz Meck
BOKELA Ingenieurgesellschaft für Mechanische Verfahrenstechnik mbH
Gottesauer Straße 28
D-76131 Karlsruhe
Germany

Dr. Harald Anlauf
Bernhard Hoffner
Institut für Mechanische Verfahrenstechnik und Mechanik (MVM)
Universität Karlsruhe (TH)
D-76128 Karlsruhe
Germany

Dr. Klaus Julkowski
KJJ Filter Engineering
P.O. Box 907
Coventry, CT 06238
USA

Graphics and Layout

Stefanie Groß
Am Dorfgraben 18
D-76872 Steinweiler
Germany

For best benefit
to Min Cheng

Reinhard [illegible]

BR 17/04/02

Our partners and clients in more than 30 countries are frequently confronted with specific 'Solid/Liquid Separation' process terms and definitions they are not used to in their normal way of doing business. Very often they are looking for a reference book with explanations and interpretations that are readily understandable. The present BOKELA SLS LEXICON aims to fulfil this need. It makes use of the definitions, terminology and concepts of the Karlsruhe School for 'Solid/Liquid Separation' (for short: SLS), as they are widely used as a standard and are meanwhile accepted to a large degree in the scientific community.

SLS forms an unique discipline with regard to process technological, equipment and scientific know-how. As a cross-sectional unit operation it is ubiquitous in all segments of the process industry as well as in the environmental protection. Hence its specialised and highly diversified terminology needs to be made accessible.

For more than 20 years, BOKELA's experts have been strong exponents of the so-called 'Karlsruhe School for Mechanical Separation Technologies', which in turn is the result of an unique, very intensive and synergetic cooperation between the different departments of the 'Institute of Mechanical Process Engineering and Mechanics of the University Fridericana of Karlsruhe' on the one side and BOKELA engineers on the other. Here we pay tribute to Professor Dr. Werner Stahl as the fountain well and spiritual rector of the Karlsruhe School.

BOKELA has been active in the process industry for more than 15 years with high expertise services, R & D, and innovative separation equipment. We are also recognised as equipment supplier and designer of comlete SLS systems. Evolving from a typical start-up company, BOKELA has had company growth that culminated in an award in 1999 by the German President Roman Herzog for a science/technology product. Nowadays, the enterprise is considered a high tech group with global reach and following clearly defined goals.

Our mission: To optimise the value of our clients through competence, creativity, reliability, speed and global coverage.

Our vision: To advance solid/liquid-separation through leadership in technology and marketing.

Driven by these goals, our experts are looking for both the ordinary and the more demanding challenges to en-

gineer them into tailor-made solutions.

The activities of numberless projects-starting with basic lab tests to market analyses and marketing plans for our clients on to the design of turnkey systems based on an unique process philosophy - give us the opportunity to increase our competence and to collect important practical experience on an on-going basis. At the same time, our engineers form trustful and valuable relationships with men and women at customers' sites all over the world. These experiences and relationships, but also the continual co-operation with our academic colleagues, are foundations for being a comprehensive innovation power. Backed by this innovation power we are convinced to keep on our responsibility for our clients and partners, but especially also for our employees.

The wonderful team atmosphere - the 'Spirit of BOKELA' - creates superior performances. And we at BOKELA are proud to be part of this team where everyone is taking care of each other. This is not only a social responsibility, but also an obligation to perform at the highest quality level. Our quality policy is backed by ISO 9001 and proclaims:

BOKELA approaches solid/liquid separation as a process technology in a comprehensive form. Integrated Engineering is the basis for the most suitable, tailor-made solutions for our clients, combining in-depth know how, wide ranging experience, and a thorough analysis of the components of an application, its pocess conditions and objectives. From this point of view, we are guided in all of our activities by the following motto:

We, the BOKELA-Team, strive to provide our clients products and services of the highest value and the best quality based on the latest state of technology. We are committed to have long-term, trusting and constructive relationships with our cllients, partners and suppliers.

Karlsruhe, April 2001

BOKELA GmbH

Dr.-Ing. Reinhard Bott

Dr.-Ing. Thomas Langeloh

Abrasion

Wear and tear on walls or pipes inside of equipment due to friction of moving solid particles; specially in areas where flow direction changes or velocity increases. Reduction is possible through smooth transitions or the application of a protective coating. The protective layer on the blades of a ➔ ***conveyor screw*** is called armoring.

Absolute Filtration

means that a specified minimal particle size has to be retained with absolute certainty.

Absorption

The transport of gases by ➔ ***diffusion*** into a condensed phase (i.e. a liquid or a solid) and forming homogeneous solutions. The gas-specific equilibrium concentration as determined by temperature and pressure limit the amount absorbed.

ACTD-Testdust

Common test material for the evaluation of a ➔ ***filter media***. **A**rizona **C**ontrol **T**est **D**ust is available in two relatively narrow fractions, i.e. Fine (0 - 80 μm) and Coarse (0 - 200 μm).

Activated Carbon (Charcoal)

Activated carbon is made from wood, peat, hardened coal or fruit shells by carbonization in an oven. They are highly porous with a large ➔ ***specific surface area*** and have excellent ➔ ***adsorption*** properties. Activated carbon is used for liquids as a ➔ ***filter media*** in ➔ ***deep bed filtration*** for clarification, decolorization, and for flavor adjustment.

Additives

Foreign matter added to a partially demoistured ➔ ***bulk***, in order to maintain a specified mean ➔ ***product moisture***. One employs this method e.g. in the processing of municipal sewage that must be deposited in dry form, however, this method may lead to a technical success but is not desirable because additional material is needed for deposition.

Adhesion

Adherence of solids caused by molecular attraction forces. Originates from ➔ ***adsorption***.

Adhesional Friction

➔ ***Friction***

Adhesive Liquid

The liquid portion in the bulk of a suspension that is still bound by ➔ ***adhesion*** after mechanical de-moisturing. The adsorbed quantity is a function of the solids specific surface, the type of liquid and its structure, respectively, reaching up to 10 molecule diameters, i.e. about 30×10^{-8} cm. This liquid cannot be removed by mechanical means.

Adsorbates

➔ *Adsorption*

Adsorption

An enrichment of gases and dissolved substances (➔ ***adsorbates***) due to molecular forces (➔ ***Van-der-Waals forces***, ➔ ***electrostatic forces***) at phase boundaries, e.g. a solid surface (adsorbens) or a liquid surface. The larger the interface the more can be adsorbed. The adsorption is normally limited to a mono-molecular surface layer. An adsorption process is usually accompanied by a release of energy (adsorption energy). The so-called adsorption isotherm correlates the concentration of adsorbate in the fluid around the boundary with the amount already adsorbed.

Agglomerate

Binding of solid particles by adhesive forces which in turn can be generated by ➔ ***Van-der-Waals forces***, ➔ ***hydrogen bridges***, or ➔ ***cross-linking polymerics***, such as ➔ ***flocculants***. Agglomerates formed like ➔ ***flakes*** enhance the separation of solids by increasing their sedimentation velocity, and often improve the permeability of a filter cake by increasing its porosity.

Alpha-Value

Factor characterizing the filter cake permeability as derived from the ➔ ***Darcy equation***. Typical alpha values of cakes range from 10^{11} m^{-2} to 10^{16} m^{-2}. They represent an integral mean over the entire ➔ ***cake thickness***. Individual alpha values can be estimated for instance by the ➔ ***Karman & Kozeny equation***. However, it can be done quantitatively only experimentally (➔ ***Filtratest***).

Ampholyte

Chemical compound, which can react both as an acid or a base, for instance aluminum hydroxide, or amino acids.

Amphoteric

Chemical compounds behave in an amphoteric manner when they respond against stronger acids as bases, or against stronger bases as acids; e.g. oxides and hydroxides of aluminum, zinc or lead, or amino acids.

Anaphoresis

➔ *Electrophoresis*

Angle of Wetting

➔ *wetting* ➔ *contact angle*

Angular Press

Special design of a ➔ ***double belt press*** by the BELLMER company, featuring a vertical ➔ ***wedge zone*** (angle) after the horizontally arranged pre-demoisturing or ➔ ***draining zone***. Subsequently, the ➔ ***sludge*** is further demoistured by pressing and shearing between the filter belts as they are guided around rollers.

Anion

A molecule with either a single or multiple negative charge. Salts can dissociate in a solution into anions and positively charged ➔ ***cations***. In an electrolyte, anions travel to the positive terminal under the application of a direct current. Anionic ➔ ***flocculation agents*** or anionic ➔ ***tensides*** carry functional groups with a negative charge.

Apex Nozzle

Discharge outlet located at the bottom of a cyclone with either fixed or adjustable cross section. The apex nozzle generally allows a much smaller volume stream than the flow of ➔ ***fines*** that is discharged through the ➔ ***vortex finder*** at the top of the cyclone.

Archimedes Number

➔ ***Archimedes-Omega Diagram***

Archimedes-Omega Diagram

Diagram used for estimating the ➔ ***settling velocity*** of particles in ➔ ***sedimentation*** when the ➔ ***Reynolds Number*** is unknown and therefore ➔ ***Stokes' law*** can not readily be applied. The dimensionless Archimedes number (Ar) containing the particle size x on the ordinate is plotted against the Omega number (Ω) involving the settling velocity w, as follows:

$$Ar = \frac{gx^3}{\nu^2}\frac{(\rho_s - \rho_L)}{\rho_L} \qquad b = \frac{\kappa\eta_L}{2p_c A^2 \Delta p}$$

where g = earth's gravitational acceleration, x = particle diameter, ρ_s = solids density, ρ_L= liquid density, ν = kinematic viscosity. Starting with a particle diameter x, the Ar number is calculated and the Ω number is read from the diagram, from which the settling velocity can be calculated.

Armored Braid

➔ ***Lace Weave***

Asymmetrical Membrane

Type of ➔ ***membrane*** with an asymmetric pore structure across its thickness. Generally, the smallest pore size structure is facing the ➔ ***suspension*** while the larger pores face the ➔ ***permeate***. Asymmetrical membranes are preferably employed in ➔ ***ultrafiltration***. The fine pore size membrane should be minimized in thickness to reduce flow resistance; a coarse pore size membrane layer serves as mechanical support underneath.

Automatic Filter

Solid-liquid separation apparatus employing candle-shaped sieves that are cleaned in-place by periodic ➔ ***back flushing***, or with a mechanical device.

Automatic Filter Press

➔ *Press filter*

Autopress

Special plate filter press by the BHS company, featuring a compressible plate packing and hermetic closure of the system with a membrane around the plate package.

Auxiliary Layer Filtration

➔ *Precoat-Filtration*

Axial Control Head

➔ *Control Head*

Back Cloth

A robust coarse mesh cloth placed between the filter cloth and the cell to prevent deformation of the cell insert under differential pressure which otherwise can increase wear and produce irregular cake formation.

Back Flushing

Technique for the cleaning of ➔ ***filter media*** or filter section of a ➔ ***depth filter***, initiated when a preset pressure drop through the filter cake is attained.

Back Flushing Filtration

is employed in ➔ ***surface filters*** for the separation of difficult to filter materials.

BOKELA ***RSF back flush filter*** with periodically or permanently working reject shoe

A novel variant represents the patented BOKELA RSF back flush filter that is similar to a ➔ ***frame filter press*** with slot-shaped suction devices that are moved periodically through the filter chambers to back flush gel-like or soft substances from the media.

Backing Fabric

➔ ***Back Cloth***

Bacteria Retaining Test

With a bacteria-retaining test, the germ load limit of a microfiltration membrane can be established for the ➔ ***degerminating filtration***. A known area of a porous ➔ ***membrane*** is coated with a suspension of test germs and the number ratio of the organism in the starting solution and in the filtrate is determined (titer reduction). A capacity of 10^7 germs /cm^2 is the minimum for a ➔ ***filter medium*** to be called a ➔ ***sterilizing filter*** (test germ: pseudomonas diminuta).

Bag Filter

Bag filters belong to the discontinuously working ➔ ***cake filters***. A filter bag, hanging in a perforated support basket, is fed with a pressurized ➔ ***suspension*** and the ➔ ***filtrate*** is collected in the containing pressure tank from which it exits. When the bag is full with solids or if an upper pressure loss is reached the solids are often disposed of together

with the filter bag. Generally, bag filters are employed to clean up liquids with low contamination. Parallel bag filters are often set up in large units to increase ➔ ***throughput*** or to operate continually.

Barometric Leg

Attachment at continuous ➔ ***vacuum filters*** to discharge the filtrate from the vacuum system without the need of a ➔ ***filtrate pump***. It is a dip tube that when completely filled with filtrate and immersed deep enough into the filtrate pool equalizes with its suction pressure the vacuum applied on the filter. Additionally produced filtrate can therefore drain off freely through the lower outlet. The installation of a barometric leg requires a sufficient overhead height, e.g. at least 8m to produce a vacuum of 0.8 bar.

Basket Weave

➔ ***Plain Weave***

Batch Separation

➔ ***Batchwise Separation***

Batchwise Separation

Discontinuous processing of a given quantity of suspension. During the batchwise separation the individual steps of the separation process, such as ➔ ***cake formation***, ➔ ***cake washing***, or ➔ ***cake demoisturing*** are not synchronized with each other and are independently and individually adjustable. In this manner one is flexible to meet the specific requirements of the product to be separated. A continuous manner of operation can be realized by a time-delayed parallel pattern of discontinuously working machines.

Batch Process

Discontinuously operating ➔ ***batch - wise separation***.

Beaker Centrifuge

Discontinuously working ➔ ***centrifuge***, used mainly in the laboratory for ➔ ***design experiments***, for preparative purposes, or for preparing small and difficult to separate product quantities, e.g. in biotechnology. Beaker centrifuges with a horizontally rotating axis feature beakers rigidly mounted to the rotor. In those with a vertical axis, the beakers are mounted vertically on a pivoting joint at the end of the rotor, and only move under the influence of the centrifugal force into a horizontal plane. The beakers can be equipped either for ➔ ***sedimentation*** or ➔ ***filtration***. Beaker centrifuges offer only very small filter areas of a few square centimeters but can be accelerated up to 10,000g. The warming up of bucket and product caused by the air friction at high numbers of revolutions and long ➔ ***centrifugation times***, can be controlled in ➔ ***cooling centrifuges***. A special design of the beaker centrifuge is the ➔ ***long arm centrifuge***.

Belt Control

Attachment in continuous filter machines with circulating ➔ ***filter media*** to ensure straightness. It usually consists

of a sensor for registering lateral belt deflections and a deflection roller correcting the belt position. Belt controls are used in ➔ ***belt filters***, ➔ ***drum filters*** with ➔ ***leaving filter belt***, ➔ ***sieve belt presses***, and ➔ ***press filter machines***.

Belt Filter

Continuously or quasi-continuously operating, horizontal ➔ ***vacuum filter***. Belt filters are offered with filter areas ranging from $1 m^2$ up to over $100 m^2$. They occupy large floor spaces and are relatively expensive compared to other continuously working vacuum filters. They are especially suited for easy filtering products that require an intensive ➔ ***cake washing***, since the washing medium can be applied on the filter cake from the top as a pool. The field of applications of belt filters is extremely broad and reaches across many industries from chemicals to mineral processing.

Belt Filter with Reversing Vacuum Trays

➔ ***Vacuum belt filter*** with periodically interrupted ➔ ***vacuum***, in order to either advance the ➔ ***filter medium*** a step, or to retract the filtrate suction trays, that are roller-mounted below the filter medium, by one step. The ability to cut the vacuum eliminates in these filters the need for a dragging seal and a circulating carrier belt. However, they consume more energy due to the periodic cell ventilation.

Belt Press

➔ ***Double-Belt Press***

Bentonites

are clay minerals (hydrous silicate of Mg and Al; main constituent montmorillonite) with a high capacity for swelling and ➔ ***adsorption***. Bentonites are employed among other uses in the clarification of beverages as well as the de-colorization of oils and fats from animal and vegetable sources.

Beta Value

The Beta value is a frequently used term for the ➔ ***filter cloth resistance***. It can be established on the basis of permeation experiments with a particle-free liquid, or directly from a ➔ ***filter experiment*** via the ➔ ***Darcy Equation***.

Binding

Type of crossing of weft and warp threads in a ➔ ***weave***. One classifies three basic bindings: ➔ ***linen or plain weave***, ➔ ***twill weave*** and ➔ ***satin weave***.

Blade

➔ ***Knife***

Blaine Number

measures according to the Blaine procedure the volume- or mass- based surface of a granular substance. The principle of the Blaine measurement is based on the gas permeation of a specified tablet manufactured of the solid to be investigated, and the evaluation of pressure-dependent gas volume flow according to the

→ ***Carman & Kozeny-equation***. The density of the solids, the → ***porosity*** of the briquette, the pressure applied, the volumetric flow rate of the gas, and its → ***viscosity*** all have to be known. It is important to recognize that with this measurement only the outer, wetted surface of the particles is taken into account, and that the numerical value obtained has to be understood less as an absolute value but more for comparing different kinds of solids. In mineral processing, for example, the Blaine Number correlates with the filterability of a suspension. Thus a Blaine number of 2000 cm^{-1} characterizes a product principally well suited for vacuum filtration.

Blinding

The progressive clogging of the → ***pores*** in a → ***filter medium*** with particles from the → ***suspension***. The free cross-sectional flow area decreases through particles penetrating the pore structure of the filter medium, and consequently its → ***pressure loss*** increases. If a critical blockage value is exceeded and the filter medium cannot be regenerated it has to be replaced.

Blockage

Clogging of the → ***pores*** of a → ***filter medium*** by particles that penetrate out of the → ***suspension*** into its structure.

Body-Feed Filtration

Special type of → ***cake filtration*** utilizing a filter aid that is admixed to the → ***suspension*** to be processed to improve its filterability. Generally, they consist of fibrous or granular substances, which open up the pore structure of the developing → ***filter cake*** and thus increase its → ***permeability***. It can also be a coarse fraction of the solids already in the suspension called a supporting grain. Body-feed filtration is an alternative to → ***precoat filtration***; it is employed in vacuum as well as in → ***overpressure filters*** and → ***press filters***.

Bond Curve

Special interpretation of the → ***capillary curve***, frequently applied in centrifugation to characterize the → ***product moisture*** that can be attained when the forces are at equilibrium. The plot of a Bond curve shows the moisture, expressed as the → ***saturation degree*** or the → ***residual moisture***, against the → ***Bond number***, which is the ratio of the respective forces expelling and retaining the pore liquid. A Bond curve falls into four characteristic sections. The first describes the conditions before reaching the → ***capillary entry pressure***, the second the region where the capillary rise still exists, the third is the so-called → ***Bond plateau***, and the last refers to the → ***demoisturing of interstitial liquid***.

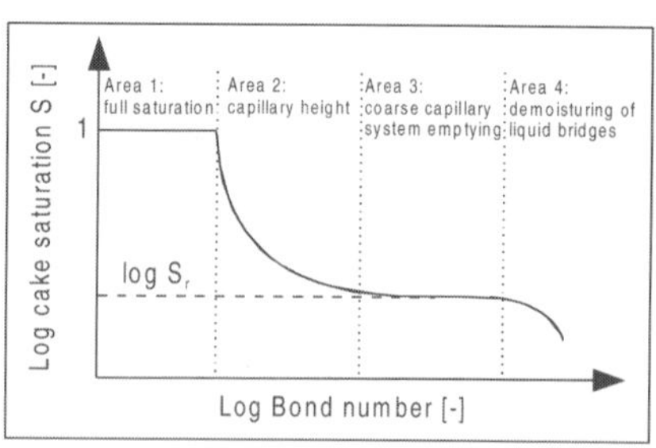

Bond Curve

Bond Number

A non-dimensional characteristic number in centrifugation describing the ratio of the forces expelling and retaining the pore liquid, respectively. The Bond number is displayed together with the ➔ ***residual moisture*** in the ➔ ***Bond curve***. Different Bond numbers Bo can be defined for the ➔ ***coarse capillary system*** and for the region of the ➔ ***interstitial water***.

Bo_1 is valid for the region of the coarse capillary liquid:

$$Bo_1 = \frac{d_h h_c \rho_L g C}{\gamma_L \cos\delta}$$

d_h ➔ ***hydraulic pore diameter***, h_c ➔ ***cake thickness***, ρ_L ➔ ***liquid density***, g ➔ ***earth's acceleration***, C ➔ ***C-value***, γ_L ➔ ***surface tension*** and δ ➔ ***wetting angle***.

In the region of the ➔ ***interstitial liquid*** the $Bond_2$ number is valid:

$$Bo_2 = \frac{d_h^2 \rho_L g C}{\gamma_L \cos\delta}$$

Bond Plateau

Special section of the ➔ ***Bond curve*** where the freely flowing capillary liquid in coarse pores has been completely removed from the ➔ ***bulk*** whereas ➔ interstitial liquid, ➔ ***adhesive liquid***, and ➔ ***inner liquid*** still are remaining. Depending on the ➔ ***surface roughness*** of granular solids, the Bond plateau is flat for smooth particles, and it declines for rougher particles towards a lower residual moisture with increasing Bond numbers.

Boozer Filter

A high performance type of ➔ ***disc filter*** of the BOKELA company. These continuously working ➔ ***vacuum filters*** are especially suited for large throughput performance, obtained by optimizing design and process know-how. Contrary to standard disc filters the Boozer filter can rotate even at 5 rpm due to the excellent hydraulic characteristics leading to a significantly increased throughput performance. The Boozer belongs to the BOKELA rotary vacuum filter family consisting of ➔ ***disc filters***, ➔ ***drum filters*** and ➔ ***pan filters*** which are of an innovative design and represent a new generation of rotary filters.

BOKELA ***Boozer disc filter*** (L-type with 2 discs and 80 m² filter area)

Bottleneck

A narrowing in a flow conduit that limits the throughput of a separation machine. Such constrictions often result from undersized ➔ ***filtrate pipes***.

Bowl Filter

Continuously working ➔ ***vacuum filter*** with a circular, horizontally segmented ➔ ***filter surface***. The bowl filter differs from the similar looking ➔ ***table filter*** by the fact that the outer flange is firmly attached to the filter disc and rotates with it. The demoistured ➔ ***filter cake*** has to be lifted over the flange with a special ➔ ***discharge screw***. This type of solids discharge requires that a protecting product layer has to remain on the ➔ ***filter medium***. The bowl filter is specially suited for coarse crystalline and rapidly settling materials, e.g. fine sands, or aluminum hydrates.

Breathing Filter Cartridge

Special filter cartridge development by the Brieden company on the basis of a ➔ ***wedge wire filter*** whose filter openings can be widened during back flushing to remove stuck particles.

Bridge Layer

Thin layer of particles generated in the first phase of ➔ ***cake formation*** producing ➔ ***bridging*** across the pores of a ➔ ***filter medium***.

Bridging

The pore openings of many ➔ ***filter media*** used in ➔ ***cake filtration*** are generally so large that a considerable share of the particles to be separated could principally traverse through them. Therefore, depending on the ➔ ***suspension concentration*** as well as ➔ ***filtration pressure***, solid particle bridges have to be built across the pores of the filter medium in the first phase of the ➔ ***cake formation***. These bridges by themselves act then as a filter medium and can retain extensively the subsequently entering solids. Bridging thereby directly influences the rate controlling ➔ ***filter cloth resistance***.

Brownian Motion

Stochastic movement of extremely small particles in ➔ ***suspensions*** caused by impacts from surrounding liquid molecules. Brownian motion is especially of relevance in the particle size region below 1 µm. Thus for example in ➔ ***gravity sedimentation*** particles with below approx. 0.5µm diameter do not settle in water due to the Brownian motion and ➔ ***thermal convection*** but instead remain suspended.

Brutsaert Equation

Approximation for describing the relative ➔ ***liquid permeability*** $p_{rel,L}$ of ➔ ***filter cakes***, defined according to ➔ ***Wyckoff & Botset***, as a function of the relative ➔ ***saturation degree*** S. The saturation degree herein refers to the region accessible for mechanical demoisturing and therefore excludes the ➔ ***residual saturation*** S_r:

$$p_{rel,L}(S) = \left[\frac{S - S_r}{1 - S_r}\right]^n$$

The exponent n in this exponential equation depends on the product and has to be determined by measurements. At each ➔ ***saturation degree*** S the equation describes the relative portion of the total permeability of the filter cake that is available during demoisturing for liquid flow in competition to the co-currently flowing gas.

Bubble Point

Term originating from the pore size analysis of ➔ ***filter media***. The bubble point characterizes the largest ➔ ***pore*** in a filter medium. A filter medium completely wetted with a liquid of known surface tension γ_L, is subjected to a gas pressure on one side which is increased in steps. As soon as the ➔ ***capillary pressure*** p_c of the largest pore is exceeded, the first gas bubble breaks through the medium. From this bubble point pressure the diameter of a circular capillary of the same pressure can be calculated with the ➔ ***Laplace-equation***, thus defining a pore size:

$$d_{bp} = \frac{4\gamma_L}{\Delta p}$$

The measuring technique is simple, quick and physically definite and applicable in the pore size region between 0.1 and 100µm.

Bubbling Zone

Special facility in ➔ ***drum filters*** for regenerating and cleaning of the ➔ ***filter medium***. A small section in the ➔ ***control head*** of the filter at the immersion point of the ➔ ***filter cell*** is designated for air blowing at low pressure from the inside of the filter cell through the filter medium. The air blow creates strong turbulences and loosens up contaminants in the ➔ ***pores*** and on the surface of the filter medium.

Buchner Funnel

Simple ➔ ***vacuum filter*** device for laboratory use, often manufactured of china or glass, with filter areas of a few cm^2 up to several $100cm^2$. The Buchner funnel is generally circular and has a flat filter support that is covered with filter paper as a ➔ ***filter medium***. The ➔ ***suspension*** to be separated is poured from above into the open funnel and filtrates into the direction of gravity. Often a water jet pump below serves as a vacuum generator.

Bucket Centrifuge

➔ ***beaker centrifuge***

Bulk

Discrete, disperse aggregate of particles, touching bodily and forming a porous layer. A bulk can develop due to ➔ ***filtration*** or ➔ ***sedimentation***; one talks about a ➔ ***filter cake*** or a ➔ ***sediment*** depending on how they are formed.

Bulk Density

The mean specific density ρ_m of a ➔ ***bulk***, calculated from the density of the solids ρ_S and the density of the fluid ρ_L in the voids, according to their respective volume fractions (➔ ***porosity*** ε):

$$\rho_m = \rho_s(1-\varepsilon)+\rho_L\varepsilon$$

If the fluid is gaseous, the gas density can be neglected compared to the solids density and the bulk density results in:

$$\rho_m = \rho_s(1-\varepsilon)$$

Bulk Material

➔ ***Bulk***

Bulk Multilayer Filter

Special form of ➔ ***deep bed filter*** or ➔ ***packed bed filter*** where the active filter layer has at least two layers of different granularity arranged on top of each other. Generally, the liquid to be purified flows first through a coarser-grained layer and then a more finely-grained layer. To avoid mixing of the different filter layer materials during back rinsing or regenerating, a fine-grained material of high density (e.g. sand) is combined with a coarse-grained material of lower density (e.g. filter coke). An important application field for these filters is the water treatment.

Buoyancy

The *static* buoyancy of a particle immersed in a fluid acts on the fluid volume that is displaced by the particle in opposition to the acceleration field formed by the pressure gradients in the fluid. The *dynamic* buoyancy occurs if an asymmetric flow develops around a particle due to its shape or if the particle is rotating. It also acts due to the asymmetrical pressure distribution with a force component on its surface per-pendicular to the flow direction.

By-pass

A branching off of a partial flow from the main flow; e.g. for collecting samples or for a measurement. Undesired by-passes can appear through leaks in pipelines or after ➔ ***crack formation*** in ➔ ***filter cakes***. In the latter flows of unutilized washing liquid develop during ➔ ***cake washing*** or of wasted gas during ➔ ***cake demoisturing***. In ➔ ***vacuum filtration*** a by-pass can be employed for regulating the filter pressure, where a controlled amount of ➔ ***secondary air*** is permitted into the low-pressure zone.

Cake

➔ *Filter Cake*

Cake Demoisturing

➔ *Demoisturing*

Cake Filtration

is a surface filtration process representing the third basic filtration type, beside ➔ ***deep bed filtration*** and ➔ ***crossflow filtration***, respectively. Its objective is to retain the solids in a ➔ ***suspension*** as a ➔ ***bulk*** on the upper side of a ➔ ***filter medium***. Both liquid or solids can principally be the desired product. Cake filtration requires a certain critical solids concentration in the feed to build a ➔ ***bridge layer*** across the ➔ ***pores*** of the ➔ ***filter medium***, which then acts as the active filter medium. The driving potential for the cake filtration can be a gas differential pressure, a mechanical or hydraulic pressing power, or a centrifugal pressure. Hence the spectrum of cake filtration equipment is wide. The cake filtration allows ➔ ***washing*** of the ➔ ***bulk*** and its mechanical demoisturing after the cake formation process. The formation of filter cakes can be described by the ➔ ***cake formation equation***.

Cake Formation Angle

Angular sector available for cake formation in a ➔ ***drum filter***, ➔ ***disc filter***, or ➔ ***table filter***. The cake formation angle α_1 is connected by the number of revolutions n with the cake formation time t_1 as follows:

$$t_1 = \frac{\alpha_1}{360°} \frac{1}{n}$$

A cake formation angle starts the earliest at the position where the ➔ ***filter cell*** is completely immersed into the ➔ ***suspension***, and it ends obviously where the cell emerges out of the suspension.

Cake Formation Equation

Relation for the description of the filter cake formation derived by combining the ➔ ***Darcy equation*** with a mass balance. For constant filtration pressure the following expression applies:

$$h_c = \sqrt{\left(\frac{R_m}{r_c}\right)^2 + \frac{2\kappa \Delta p t_1}{r_c \eta_L}} - \frac{R_m}{r_c}$$

For constant feed flow rate it has the form below:

$$\Delta p(t_1) = \frac{\dot{V}_L \eta_L}{A}\left(\frac{r_c \kappa \dot{V}_L t_1}{A} + R_m\right)$$

h_c = cake thickness, r_c = specific cake resistance, R_m = filter cloth resistance, κ = ➔ ***Kappa-Factor***, Δp = pressure difference, t_1 = cake formation time, A = filter area, η_L = dynamic viscosity

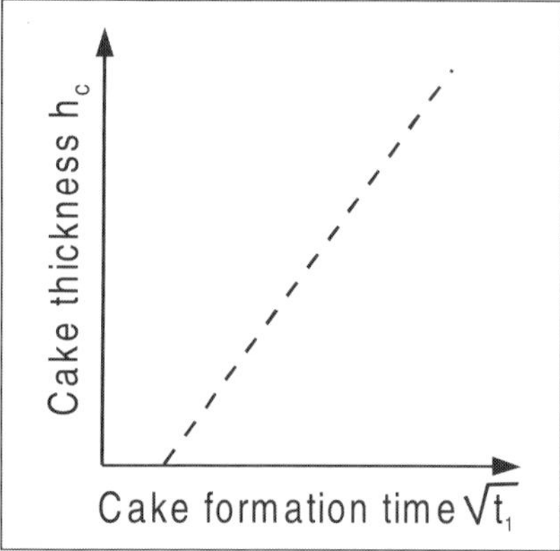

Cake thickness as function of cake formation time

Cake Permeability

The cake permeability p_c is the reciprocal of the ➔ ***cake resistance*** r_c:

$$p_c = \frac{1}{r_c}$$

The cake permeability is quoted in the unit ➔ ***Darcy*** (cm^2). It can be directly determined by means of the ➔ ***t/V=f(V)-method*** from a filter experiment.

Cake Resistance

Specific permeation resistance r_c of the ➔ ***filter cake***. It is independent of the cake thickness and can be directly determined via ➔ ***Darcy's law*** by flowing a particle-free liquid through two filter cakes of different heights at a predetermined pressure and measuring the rate. From the resulting two equations the filter cloth resistance can be eliminated. Normally, the cake resistance is determined directly by the ➔ ***t/V=f(V)-method***.

Cake thickness

as produced by ➔ ***cake filtration*** equipment can range from a few mm up to 1m height. Both filterability of the ➔ ***suspension*** and design features play important parts. The cake thickness can be described with the ➔ ***cake formation equation***.

Cake Washing

Method for the removal of ➔ ***mother liquor*** and dissolved substances from a ➔ ***filter cake*** with a liquid, mostly miscible with the ➔ ***mother liquor***. One differentiates between ➔ ***displacement washing*** and ➔ ***dilution washing***; the latter involves re-suspending of the filter cake in the wash liquor. Quality criteria for displacement washing are a high ➔ ***wash degree***, and low consumption of wash liquid which can be expressed by the ➔ ***wash ratio***.

Calendering

is a thermal-mechanical surface treatment process for smoothening of woven surfaces. A ➔ ***weave*** made of thermally workable material is treated with pressure in a calender roller to give a smooth surface to a ➔ ***filter medium***, which for example facilitates cake discharge. It should be noted that the resulting ➔ ***pore size*** of the filter surface is decreased by calendering.

Candle Filter

Discontinuously working filter equipment designed as either ➔ ***cake filter*** or ➔ ***deep bed filter***. Often, cylindrically shaped filter elements are manifolded together in large number in a pressure tank to house an economically reasonable ➔ ***filter area*** in a given vessel volume. Candles covered with a ➔ ***filter cloth*** are applied in conventional ➔ ***cake filtration*** or in ➔ ***precoat filtration***. They can discharge a dry cake after gas pressure demoisturing or a highly thickened ➔ ***suspension*** after dropping it in the surrounding heel (e.g. ➔ ***Fundabac filter***). If candle filters are employed as deep bed filters they serve the purification of liquids polluted with small amounts of contaminants. So-called ➔ ***rolled candles*** and elements made from sintered materials are also employed in candle filters.

Candle Press Filter

Special type of ➔ ***membrane filter press***, at which the press membrane is arranged in a circular, cylindrical manner around a ➔ ***filter candle***. This special design variant allows pressing forces of considerably exceeding 100bar; it is used in extremely difficult to filter products, e.g. in the field of biotechnology.

C.A.P.

The ***C***ontinuous ***A***rea ***P***ress of the BOKELA company represents a continuously working post-demoisturing machine for ➔ ***compressible*** ➔ ***sludges*** that are pre-demoistured. The feed after spreading to uniform height in the CAP is continuously pressed out between two sieve belts on a press roller. At area pressures of up to 30bar it is used for example to produce paper slurries.

Capillarity

The behavior of liquids caused by ➔ ***interfacial tension***.

Capillary

(fr. L capillaris: hair). Fine pore channel in a porous ➔ ***bulk*** or in a ➔ ***filter medium***.

Capillary Belt Filter

Filter machine in which the liquid flows out of a ➔ ***bulk material*** through the openings of a ➔ ***vibrating screen*** where it is received by an absorbing belt underneath, circulating against the solids transport direction. The absorbent medium is then squeezed between two press rollers and available again for a liquid intake upon return.

Capillary Condensation

denotes liquefaction of vapors in fine ➔ ***pores*** (➔ ***capillaries***) of a porous solid material. There is a strong physical relationship between the ➔ ***capillary pressure*** in such a pore and the relative humidity in the surrounding gas phase.

Capillary Diameter

is an ➔ ***equivalent diameter*** d_{cap} for circular and cylindrical ➔ ***capillaries*** as determined by the ➔ ***Laplace equation***

from a ➔ ***capillary pressure*** p_c:

$$d_{cap} = \frac{4\gamma_L \cos\delta}{p_c}$$

γ = interfacial tension, δ = wetting angle. This value is derived from a ➔ ***Bubble Point Test***.

Capillary Entry Pressure

As capillary pressure is indirectly proportional to the diameter of a ➔ ***capillary***, the pressure of the largest capillary in the ➔ ***bulk*** that bubbles first is called the capillary entry pressure. This is the minimum gas pressure to be applied from the outside to demoisture the largest capillaries.

Capillary Liquid

denotes in general the liquid contained in the ➔ ***pores*** of the ➔ ***bulk*** which are hydraulically connected to each other and thus accessible to mechanical demoisturing (➔ ***coarse capillary liquid***).

Capillary Module

➔ ***Hollow Fiber Module***

Capillary Pressure

is the ➔ ***pressure difference*** across a curved gas-liquid phase boundary, which is compensated by the ➔ ***interfacial tension***. The capillary pressure in the ➔ ***pores*** of a ➔ ***filter cake*** holds the liquid in it. A capillary pressure can have positive or negative values. Accordingly, a capillary pressure is positive if a lower pressure exists in the phase with the larger density. Therefore, the capillary pressure of a liquid droplet in a gaseous environ-ment is negative, whereas the capillary pressure of a gas bubble in a liquid would be positive.

Capillary Pressure Curve

is a function that describes the ➔ ***capillary pressure distribution*** within a ➔ ***bulk*** due to the underlying pore size distribution via ➔ ***capillary pressure*** against ➔ ***saturation degree***. This function defines what minimum saturation degree can be reached at equilibrium for each pressure difference that is constantly applied on the ➔ ***bulk*** from the outside.

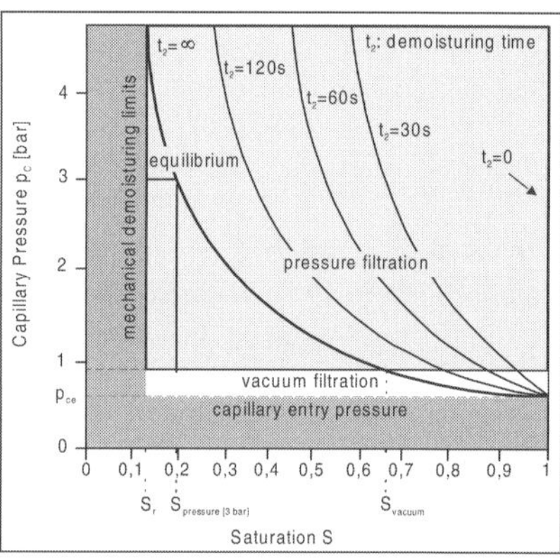

Capillary Pressure Curve

Special values of the capillary pressure curve are the ➔ ***capillary entry pressure*** and the ➔ ***remanent saturation***. The capillary pressure curve can be determined for an applied gas difference pressure or for a centrifugal field; both measurements results should be theoretically the same. The capillary pressure curve in homogeneous

➔ ***bulks*** does not depend on the physical height of the ➔ ***bulk***.

Capillary Pressure Distribution

➔ ***Capillary Pressure Curve***

Capillary Rise

Height of a liquid column in a wetted ➔ ***capillary*** or in a ➔ ***bulk***, adjusting itself against the hydrostatic pressure in equilibrium:

$$h_{cap} = \frac{p_c}{\rho_L g} = \frac{4\gamma_L \cos\delta}{\rho_L g d_{cap}}$$

γ = interfacial tension, δ = wetting angle, p_c = capillary pressure, g = gravitational acceleration, ρ_L = liquid density, d_{cap} = capillary diameter

Capillary Suction Draught

➔ ***Capillary Pressure***

Carman & Kozeny-Equation

is used to quantify the permeation velocity v in ➔ ***bulk solids*** analogous to ➔ ***Darcy's law***, however, it explicitly describes the influences, respectively, of the porosity ε, the volume specific solids surface S_V, the fluid viscosity η_L the driving pressure difference Δp, and the filter layer thickness h_c:

$$v = \frac{1}{k(\varepsilon)} \frac{\varepsilon^3}{(1-\varepsilon)^2} \frac{1}{S_V^2} \frac{1}{\eta} \frac{\Delta p}{h_c}$$

The value of $k(\varepsilon)$ is constant at around 4 in the porosity region of $0.3 \le \varepsilon < 0.65$ in the equation. $k(\varepsilon)$ has to be determined precisely by a calibration measurement. The equation emphasizes the strong influence porosity has on the permeability of ➔ ***bulk materials***. A similarly structured relation is given by the ➔ ***Gupte-Equation***.

Cataphoresis

➔ ***Electrophoresis***

Cation

A positively charged ion in a solution (comp. ➔ ***anion***)

Caulking Thread

used for fastening the ➔ ***filter cloth*** on the cells of a ➔ ***drum filter***. For this the filter cloth is wrapped around the ➔ ***filter cell*** with the aid of the caulking thread in dovetail grooves.

Cell

➔ ***Filter Cell***

Cell Belt Filter

➔ ***Tray Belt Filter***

Cell Drum Filter

➔ ***Drum Filter***

Cell Foot

Transitional section of a ➔ ***filter cell*** in a ➔ ***disc filter*** into the ➔ ***filtrate pipe***.

Cell Insert

Structure, usually made of plastic, to

stabilize the cells interior against the acting differential pressure and to simultaneously enhance ➔ ***filtrate drainage***. Cell inserts have different geometric shapes in order to optimize mechanical stability, costs, and ➔ ***permeability***, respectively.

Cell Less Drum Filter

➔ ***Drum filter*** with a non-partitioned interior that is completely exposed to a ➔ ***vacuum***. For the cake removal a so-called control shoe with sliding seals is pushed in the removal zone against the drum's inner wall for cake removal by ➔ ***compressed air backpulse***. The ➔ ***filtrate*** in turn is withdrawn from the base of the drum's internal space by a filtrate suction pipe.

Cellulose

➔ ***Filter aid*** manufactured from coniferous or deciduous trees, or other renewable resources (e.g. ➔ ***Rebecel***) by cooking, bleaching, and grinding. Cellulose is often offered as mixture with ➔ ***diatomaceous earth***.

Celtic-Weave

➔ Plain Weave

Censor

Special development of a ➔ ***decanter centrifuge*** by the KHD company for separating solids with different densities. The centrifuge rotor has a double conical shaped drum and a transporting conveyor screw, that transports from middle in both directions. This screw turns with a differential speed relative to the drum. The separation conditions have to be adjusted in such a manner that the heavy solids settling out of the ➔ ***suspension*** can be conveyed and discharged in one direction while light solids floating up in the suspension liquid in the other direction. The main field of application is for sorting of plastics.

Centrate

Cleared liquid evolving from a ➔ ***centrifuge***.

Centridry

Process from the KHD company for complete drying of sludges. A combination of ➔ ***centripress*** ➔ ***decanter centrifuge***, conceived for maximum mechanical demoisturing, and a thermal flow dryer. A predemoistured sludge ejected from the centrifuge discharge at a high kinetic energy is finely dispersed in a hot gas stream and thermally dryed in an extremely short period of time. The main applications are waste sludges.

Centrifugal Extractor

Special design of a ➔ ***disc stack separator***, used for the separation of two immiscible liquids with different densities. The heavy phase is discharged with a ➔ ***peeling disc*** by adding extraction agents into the mixing chamber. A complete and spontaneous mixing of the extraction agent is achieved through the peeling disc.

Centrifugal Filter

Filter centrifuge with a perforated ➔ ***drum*** that is covered with a ➔ ***filter medium***. Centrifugal filters can demoisture cake solids to a great extent by removal of the fluids that are held by capillary forces, and can produce a free-flowing solid product.

Centrifugal force

A particle in circular motion strives to leave the circle in the direction of the tangent to its path according to the law of inertia. A force has to be exercised on it constantly, pointing to the center of the circle, called the centripetal force. According to the counter effect principle the centrifugal force corresponds to this as an oppositely directed, equal-sized force. It represents the inertial resistance with which the moving particle opposes the constant change of the direction of its path.

Centrifugal Machine

➔ ***Centrifuge***

Centrifugal Mixer

A rotating mixing chamber fed centrally through concentric pipes with liquid components and equipped with a stationary ➔ ***peeling disc*** for the discharge of the mixed liquid.

Centrifugal Time

Residence time of solids to be separated out of a ➔ ***suspension*** in the process chamber of a ➔ ***centrifuge***.

Centrifugal Value

➔ ***C-value***

Centrifugation

Solid-liquid or liquid-liquid separation process in a ➔ ***centrifuge***.

Centrifugation Superimposed by Overpressure

Design of ➔ ***siphon peeler centrifuges*** (Krauss-Maffei company) as well as in ➔ ***inverting filter centrifuges*** (Heinkel company), where the interior space of the drum is put under a gas overpressure, so that during the ➔ ***filtration*** not only the centrifugal pressure but also a gas pressure is utilized. Beneficial for products with a high ➔ ***capillary rise*** of the cake water. Overpressures of up to 6 bar are applied.

Centrifuge

Rotating separation apparatus employing centrifugal forces as driving potential. Mass forces are produced that act both on the particles to be separated as well as on the liquid contained in the centrifuge. Centrifuges are used for ➔ ***sedimentation*** and ➔ ***filtration***. The range of ➔ ***C-values*** in commercial centrifuges starts in the low 100's and can go up to several 10,000.

Centrifuge Rotor

The machine component of a ➔ ***centrifuge*** in which a separation process is performed at high speeds by ➔ ***sedimentation*** or ➔ ***filtration***.

Centrifuge Value

➔ *C-value*

Centripetal Force

➔ *Centrifugal Force*

Centripetal Pump

➔ *Peeling Disc*

Centripress

➔ ***Decanter centrifuge*** developed by the KHD company for maximal sludge demoisturing. A high pressing of the sludge is achieved in this centrifuge by specially high damming of the sludge bed and a special design of the ➔ ***conveyor screw***. This mode of operating a decanter centrifuge necessitates producing extremely high ➔ ***torques***, which in turn requires special gears and drives. It also applies for comparable machines of other manufacturers. Main application for this type of centrifuge is processing of municipal sewage sludge.

Centrisizer

➔ ***Decanter centrifuge*** developed by the KHD company for ➔ ***classifying*** a ➔ ***suspension*** into two grain fractions. This centrifuge discharges both fractions in sludge forms through nozzles to minimize any turbulence in the processing space, which would otherwise reduce the ➔ ***separation selectivity***.

Centritest

➔ ***Laboratory beaker centrifuge*** of the BOKELA company for filtration and sedimentation experiments. It is a ➔ ***quick start centrifuge*** for C-values between 0 and 6,000g with a rotor diameter of 580mm. The kinetic course as well as the equilibrium states in separations can be monitored with telemetric data transfer and a computer-aided data analysis. Optionally, the centrifugation can be superimposed by a pressure filtration, either with a gas pressure difference of up to 7 bar or by a steam pressure filtration. As a ➔ ***quick start centrifuge*** the Centritest is also well suited for the scale-up of continuous working screen centrifuges with a very short residence time of the product.

BOKELA ***Centritest***

Ceramic Disc Filter

A continuously operating vacuum filter by the OUTOMEC company based on a ➔ ***disc filter***. In place of conventional ➔ ***filter cells*** the filter elements are ceramic plates filtering on both sides. The microporous filter plates have approx. 1 µm pore size, are each several mm thick, and therefore impose a high flow resistance that controls the filtration process. The minute pore width is ad-

vantageous for producing ➔ ***filtrates*** free of particles and without gas flow in the vacuum region. The ➔ ***capillary pressure*** of the ➔ ***hydrophilic*** ceramic media is larger than the applied ➔ ***pressure difference***, so that the ➔ ***pores*** remain always wetted.

Chain Assisted Cake Removal

Special type of filter cake removal in ➔ ***drum filters*** where parallel running chains or strings (string assisted cake removal) are guided around the filter drum and are led away from the drum in the cake removal sector. Following a sharp deflection by a roller system the chains are routed back to the drum. The ➔ ***filter cake*** consisting of mostly fibrous solid particles builds up around the chains. Therefore the cake is picked up in the removal sector of the drum by the chains like a ➔ ***fleece*** and is thrown off at the sharp chain deflection.

Chamber Filter Press

The most common design variant of the ➔ ***filter press*** filter where two neighboring filter plates with a one-sided cavity face each other to form a filter chamber. The ➔ ***filter medium*** is stretched over each plate and is pressed into the chamber during the process. The discharge of the cake from chamber filter presses can be executed far more beneficially and simpler than from ➔ frame filter presses. However, since the cake discharges due to its weight by dropping out of the chambers, a certain cake thickness of several centimeters is required, which tends to lengthen the cycle. In addition to this disadvantage, a further problem can be that residual volume filtration will not be possible, if the suspension feed stops when the chambers filling has not been completed. In this regard the ➔ ***membrane filter*** presses are clearly at an advantage.

Channeling

Phenomenon observed during ➔ ***sedimentation*** of particles in the region of ➔ ***swarm sedimentation***. At certain concentrations (➔ ***intermediary suspensions***) hydrodynamic instabilities can occur that form particle-free liquid channels parallel to the sedimentation direction. Similar effects are known for the flow through fluidized beds.

Charcoal

➔ ***Activated Carbon***

Chemisorption

➔ ***Adsorption***

Chitosan

Food-suitable ➔ ***flocculation agent*** produced from shells of crustaceans.

Circular Thickener

➔ ***gravity thickener***

Clarifying

has the process-related objective of removing solid particles or ➔ ***colloids***, respectively, from a liquid by ➔ ***sedimentation*** or ➔ ***filtration***.

Clarifying Filtration

is defined as the complete separation of all particles from a dilute suspension (e.g. beverage), most frequently done by ➔ ***deep bed filtration***.

Clarifying Separator

Special design of a ➔ ***disc stack separator*** for separating solid particles and liquids. Herein the ➔ ***disc package*** does not contain any ➔ ***rising channels***, as in the disc stack separators, which are conceived for the separation of two non-miscible liquids with different densities and potentially solids, too. Clarifying separators are designed with respect to the feed solids content as ➔ ***nozzle-type separators*** or as ➔ ***self-cleaning separators***. They are able to reach ➔ ***C-values*** of more than 15,000 and have ➔ ***equivalent clarifying*** areas of up to 100,000 m^2. Clarifying Separators are employed for the separation of extremely small particles down to the sub-µm region.

Classifying

Separation of a particle collective into ➔ ***fractions*** of different ➔ ***particle sizes***.

Clear Liquid

commonly refers to the overflow of ➔ ***gravity thickeners*** with as few particles as possible.

Clear Liquid Zone

Nearly particle-free zone developing in the upper section of a sedimentation tank (➔ ***gravity thickener***), from where the separated liquid is taken off with an overflow.

Clear Run

Phase in the ➔ ***cake filtration*** process after the formation of solid bridges over the ➔ ***pores*** of the ➔ ***filter medium*** when the ➔ ***solids loss*** ends and a clear ➔ ***filtrate*** begins to flow.

Clogging Layer

Phenomenon in the ➔ ***cake filtration*** under the influence of gravitational or centrifugal forces, observed as a partition of the ➔ ***suspension*** with respect to its ➔ ***particle size***. The larger solid particles settle quickly on the ➔ ***filter medium*** according to ➔ ***Stokes' law***, whereas the smaller particles are deposited later on the cake surface forming a so-called clogging layer. Often, this clogging layer is highly impermeable and it will then increase the ➔ ***capillary entry pressure*** and thus the ➔ ***residual moisture*** of the resulting ➔ ***filter cake***.

CMC

➔ ***Critical Micelle Concentration***
➔ ***Micelles***

Coagulation

➔ ***Agglomeration*** of fine-grained particles by the ➔ ***destabilization*** of a ➔ ***suspension***, due to a change in the ionic make-up, which leads to a dominance of the attracting ➔ ***Van-der-Waals forces***.

Coarse Capillary

→ *Coarse Capillary Liquid*

Coarse Capillary Liquid

is defined as the major portion of liquid in a saturated → ***bulk*** that is held between the particles in hydraulically inter-connected voids. The coarse capillary liquid is readily accessible to mechanical demoisturing by gas differential or centrifugal pressure. Beside the coarse capillary system exists also a → ***fine capillary system***.

Coarse Capillary System

→ *Coarse Capillary Liquid*

Coarse Material

When a → ***suspension*** or a → ***bulk material*** displaying a → ***particle size distribution*** is separated into two → ***fractions*** of different grain size, the fraction containing the larger ones is called the coarse material.

Coarse Screen

Equipment employed in sewage treatment for the screening of coarse particles in the cm-size region, such as paper, wood, plant refuse, plastics. The screen is periodically cleaned off the accumulated debris with an automatic, comb-like device.

Crimping

Soaking of weaves in boiling water, applied as pre-treatment, e.g. with cotton fabrics to render them resistant to shrinking.

Co-current Flow Decanter

Special flow distribution of the centrate in a → ***decanter centrifuge***. The feed → ***suspension*** enters at the cylindrical end of the centrifuge, so that solids and → ***centrate*** can move together towards the conical end. From there the centrate is returned by channels attached along the structure of the → ***transport screw*** for discharge from the cylindrical end. The purpose for this design is to disturb the settling process of the solids as little as possible. It affords readily a high separation degree, i.e. a very clear centrate, with solids that tend to get re-suspended easily by eddies.

Co-current Washing

is the simplest way of carrying out → ***displacement washing*** or → ***dilution washing*** where the wash liquid is brought only once into contact with the particle system to be washed. This method consumes more washing liquid than → ***countercurrent washing***. The latter, however, is not feasible with all types of separation equipment. In → ***centrifuges*** for example only the co-current washing method can be employed due to a lack of possibilities for segregating the centrate.

Coiled Candle

→ ***string-wound cartridge***

Colloid

So-called colloidal disperse systems are aggregates of molecules, comprised

10^3 up to 10^9 molecules; they would have a diameter of 10^{-7} to 10^{-4} cm if they assumed a spherical shape. Colloids assume an intermediate position between the molecular disperse and the coarsely disperse systems. They are difficult to separate in stable ➔ ***dispersions*** and cause turbidity in a separated ➔ ***clear liquid***.

Combination Arrangement

In-series arranged solid-liquid separation machines for dividing the separation process into sections of different liquid content. A typical combination may consist of a pre-thickener (e.g. static ➔ ***circular thickener***), followed by a mechanical demoisturing apparatus (e.g.➔ ***vacuum drum filter***), and last, a thermal drying step (e.g. spin flash-dryer). Through proper serial combination of separation equipment a desired separation can be technically realized and at the same time economically optimized.

Composite Membrane

➔ ***Membrane*** consisting of at least two different materials which are solidly attached to each other. Composite membranes are employed especially in ➔ ***ultrafiltration*** which demands extremely small membrane pores for the material retention. They consist of a fine porous, very thin cover layer with high flow resistance and a large-pored, mechanically stable support layer underneath. ➔ ***Weaves*** and ➔ ***fleeces*** are often used as a support layer.

Compressed Air Blow Back

The energy source for detaching a ➔ ***filter cake*** from a ➔ ***filter cloth*** at the ➔ ***solids discharge*** of ➔ ***drum***, ➔ ***disc***, ➔ ***candle*** and ➔ ***leaf filters***. After demoisturing a sudden increase of pressure is generated on the filtrate side behind the ➔ ***filter medium***, in order to break the cake up or cast it away. A critical point with pressurized air blow back is the potential for filtrate residues remaining in the cloth or in the ➔ ***filter cell***, which can consequently cause ➔ ***re-moisturizing*** of the filter cake.

Compressibility

The property of a ➔ ***bulk*** to decrease its void space due to a load acting from the outside. The dimensionless compressibility degree U is used to quantify compressibility. U relates the difference between the original layer thickness L_1 and a present layer thickness L to the maximum compression (L_1 -L_∞), which in turn is the difference between original layer thickness L_1 and the layer thickness reachable at equilibrium L_∞:

$$U = \frac{L_1 - L}{L_1 - L_\infty}$$

Compression

Process employed in solid-liquid separation technology for the ➔ ***demoisturing*** of ➔ ***bulk materials***, which do not have a rigid pore matrix. The liquid is squeezed to the outside by decreasing the pore volume, so that the ➔ ***dry substance*** content increases. Compression is technically preferably

achieved with the aid of ➔ ***press filters***. During ➔ ***sedimentation***, the developing ➔ ***bulk*** is furthermore subject to compression by the acting mass forces. Compression does not only occur with elastic but also with rigid particles. There are four different modes of compression: first, the particles approach each other; secondly, they glide past each other; third, further compression can be achieved to a certain extent by rearranging the particles, and fourth, a particle fracture occurs.

Compression Layer

Term used in the field of ➔ ***sedimentation***. The compression layer in a ➔ ***thickener*** defines the region where the settled particles at the bottom of the tank approach each other to such an extent that they are capable of transferring mechanical forces among each other. The sediment in this region is compressed due to its weight. The higher the layer and the longer the time spent the stronger the compression.

Compression Work

Energy W, which has to be applied for the compression of a gas with the volume V_{g1} (air) by the pressure p_{g1}, in order to generate and maintain a pressure difference ($\Delta p = p_{g2} - p_{g1}$), necessary for vacuum or overpressure filtration. It is calculated for an isentropic change of state with the isentropic coefficient κ:

$$W = p_{g1} V_{g1} \frac{\kappa}{\kappa - 1} \left[1 - \left[\frac{p_{g2}}{p_{g1}} \right]^{\frac{\kappa-1}{\kappa}} \right]$$

Concentrate

Term used especially in the region of the ➔ ***micro-*** and ➔ ***crossflow filtration*** for the resulting ➔ ***suspension***, after ➔***permeate*** is drained off through the ➔ ***membrane***. The concentrate is in any case still a free-flowing ➔ ***sludge***.

Concentration

➔ ***Solids Content***

Condensate Front

Phenomenom at the ➔ ***Steam Pressure Filtration*** on ➔ ***Hi-Bar Filters*** which provides for excellent filter cake demoisturing and filter cake washing. A condensate front is formed when a "cold" and saturated ➔ ***filter cake*** enters the hot atmosphere of overheated or saturated steam in a ➔ ***steam cabin***. Then, the steam condenses on the surface of the cake and the condensation enthalpy heats the cake surface up to the condensate temperature. While the pressurised steam forces the ➔ ***mother liquor*** and the condensate through the cake, cold regions of the filter cake come in contact with steam and further condensate is formed. This leads to a sharply defined and evenly developed condensate front which moves through the cake as a homogenous conden-sate layer preventing a ➔ ***fingering***. Thereby,

the mother liquor is displaced and the cake is completely heated up to the temperature of the condensate. This mechanism combines heat and mass transfer between filter cake and steam.

Concentration Polarization

Term out of the field of ➔ ***crossflow filtration***. At the permeation of the filter a convective transport of the materials to be retained develops in the direction of the membrane due to the ➔ ***pressure difference*** across the membrane. In the course of this developing concentration gradient, a diffusive back transport into the core flow results. In the stationary case an equilibrium between these two processes is reached.

Conditioning

The changing of the properties of a ➔ ***suspension***, a ➔ ***sludge***, or a ➔ ***filter cake***. Suspension conditioning is defined as a pretreatment for the improvement of the separation characteristics. This can be for example a ➔ ***flocculation*** by addition of a polymer. The sludge conditioning can also include an agglomeration process for the improvement of additional demoisturing steps. Slurries or ➔ ***filter cakes*** can be conditioned with ➔ ***additives***, such as lime, to increase their dry substance content as required for landfilling, for example.

Consolidation

➔ ***Compression***

Contact Angle

The contact angle δ characterizes the ➔ ***wetting*** of solids by a liquid. One talks about wetted conditions, if the contact angle, measured always in the fluid phase with a higher density, is $< 90°$; is it $> 90°$ one talks about non-wetted conditions.

Contact Point Number

Count of contact points a particle has with neighboring particles in a ➔ ***bulk material***. In regular packages, exactly defined in geometric terms, the contact number is fixed. Thus a particle in densest possible sphere package possesses 12 contact points to its neighboring particles. The contact point number influences the ➔ ***bulk*** properties, especially in the region where the liquid is bound essentially only in the form of ➔ ***liquid bridges***. It influences the ➔ ***tensile strength*** and the ➔ ***saturation degree*** of the moist ➔ ***bulk***.

Contibac

Special-design ➔ ***candle filter*** by the DrM, Dr. Müller company for the quasi-continuous ➔ ***thickening*** of difficult to filter ➔ ***suspensions*** of catalyst slurries, dilute solids, and crystals (➔ ***Fundabac filter***).

Continuous Area Press

➔ ***C.A.P.***

Continuous Phase

In a mixture out of solids and liquid for example one of both substances re-

presents depending on the → ***concentration*** the continuous and the other the discontinuous or → ***disperse phase***. The continuous phase is connected together and encloses the discontinuous phase. In a → ***suspension*** the liquid presents the continuous phase, whereas the dispersed solids forms the discontinuous phase. By comparison at a → ***filter cake*** the connected solids structure forms the continuous phase and the liquid is distributed as a discontinuous phase in the → ***pores*** of the → ***bulk***.

Continuous Separation Apparatus

Separation apparatus which is equipped with a constant feed for the suspension and steady discharges of separated liquid and moist solids. The individual steps of the separation process therefore are synchronized in terms of transport velocity and the geometry of the apparatus and are not independently adjustable. Continuously working machines are typically employed for large mass flows and continuous production processes. → ***decanter centrifuges*** or → ***drum filters*** are examples for continuously working separation apparatus.

Control Disc

A disc made often out of plastic, employed at the → ***control head*** at → ***rotary filters***, into which the control zones are cut as slots. The control disc is stationary while the → ***filtrate pipes*** move from control zone to control zone (e.g. cake formation zone, demoisturing zone, → ***compressed air blow back*** zone, cloth cleaning zone).

Control Head

Interface between the stationary and the rotating part during the draining off of filtrate at → ***rotary filters***. A control head is divided into different control zones, in order to adjust independently and separately from each other the → ***pressure difference*** in the cake-formation and demoisturing region, as well as the → ***compressed air blow back*** for the cake discharge, and possibly a → ***bubbling zone***. Respective to their design, one can differentiate between axial and radial control heads. In the axial type the division into zones is performed by the so-called → ***control disc***. It is a plastic disc into which the slot-shaped control zones are cut. The control head sits flat with seals against the rotating counter part of the integrated → ***filtrate pipes***. In a radial control head, the rotating part moves concentrically in a stationary ring with a channel, which can be divided by so-called separation plugs into separation zones. The sealing is with gland packages. In industrial practice the axial control head is preferred due to its easier sealing and lower design complexity.

Control Valve

→ ***Control Head***

Conveyer Screw

→ ***Discharge Screw***

Conveyor Chute

Discharge device for solids (➔ ***discharge chute***), usually designed in the form of a pipe, through which the separated and demoistured solids leaves the process chamber of the separating apparatus under the influence of the gravitational force. The requirement for the functioning of a conveyor chute is a sufficient pourability of the product. If this tends to stick, then one selects a ➔***discharge screw*** instead of a conveyor chute.

Cooling Centrifuge

Discontinuously working laboratory centrifuge with integrated cooling of the beakers. Especially with high speed ➔ ***centrifuges*** in the region of several thousand g the rotor is heated by the friction of surrounding air. If a constant temperature is desired, for example due to probable product destruction otherwise, the rotor has to be cooled.

Coriolis Force

is acting during a relative movement of a particle against a rotating reference system in the centrifugal field. It is an ➔ ***inertia force*** usually possessing a negligibly small value.

Coulomb's Friction Law

➔***Friction***

Counter Flow Decanter

Conventional and simple design of a ➔ ***decanter centrifuge***. The ➔ ***suspension*** is fed approximately in the middle between the cylindrical and the conical end of the centrifuge. Solids and ➔ ***centrate*** stream then in a counter flow. While the ➔ ***transport screw*** conveys the solids into the direction of the conical end, the centrate flows in the opposite direction to the cylindrical end, where it leaves the centrifuge for example over a ➔ ***weir disc***.

Counter Ions

Layer of ions, relative strongly bound to a charged particle surface, out of the surrounding liquid with a particle surface of opposing charge. With increasing distance from the particle surface the concentration of the counter ions, compensating the surface charge, decreases. According to the ➔ ***DLVO-Theory*** the compensation of the particle charge by the counter ions determine to what extent particles are capable of agglomerating due to their attracting ➔ ***Van-der-Waals forces***.

Countercurrent Washing

Process for the removal of unwanted solutes from ➔ ***suspensions*** or ➔ ***filter cakes***, especially for saving washing liquid. The countercurrent washing can be applied in ➔ ***displacement washing*** as well as in ➔ ***dilution washing***. The fresh washing liquid is added at the end of the separation process in order to clean the already almost completely cleaned product. The washing liquid by this now already enriched with some ➔ ***mother liquor*** is then utilized for the purification of the even stronger contaminated product located further in the direction of the process beginning. Finally, the highly enriched wash liquid is dis-

charged out of the process.

Cricket-Filter

Discontinuously, in a pressure tank under overpressure working ➔ ***cake filter*** by the AMAFILTER company. The filter elements are similar to ➔ ***filter candles*** but however display a flattened shape. They thus represent a connecting link between candle filters and ➔ ***leaf filters***. Several filter elements are comprised in a register. In a pressure tank several register are arranged parallel. Cricket filters are suited for dry- and wet-discharge of a product and can be employed as a ➔ ***precoat filter***. The cricket filter is a typical representative of a ➔ ***fine filter*** for ➔ ***suspensions*** difficult to filter.

Crossflow Filtration

comprises all filtration techniques, where the ➔ ***suspension*** to be separated is lead under a pressure in parallel flow to a ➔ ***filter medium***, in order to prevent the deposition of solids on the filter medium as much as possible. Due to a pressure difference, the ➔ ***permeate*** penetrates the filter medium. The ➔ ***concentrate*** leaves the process chamber highly concentrated but still free flowing. The shear flow at ➔ ***dynamic crossflow filters*** can additionally be increased by additional stirring elements (DYNO-Filter of the BOKELA company). ➔ ***Microporous membranes*** are usually employed as filter media. The field of application for crossflow filters is vast and reaches from upgrading of sewage to product isolation at biotechnological productions. The production of extremely pure liquids (➔ ***sterile filtration***) from suspensions with often extremely low particle concentrations in the µm- and sub-µm region and poor filtration behavior is characteristic. Both ➔ ***pre-coat filtration*** and ➔ ***disc stack separators*** are in competition with crossflow filtration.

Cushion Module

Special packaging of a membrane filter medium in a micro- or ultra filtration unit. Here two membrane discs are welded together at the outer edge. Between both membrane discs a ➔ ***spacer*** is situated through which the ➔ ***permeate*** is discharged by means of a permeate drain off pipe, connected tightly with the membrane cushion. Designs with circular or rectangular membrane cushions are known. In the case of a circular membrane cushion the permeate flows to a centered discharge pipe. Several membrane cushions are combined in a pipe-shaped casing to a membrane module.

Cut Point

➔ ***Cut Size***

Cut Size

➔ ***Particle size***, which is found in equal amounts in the ➔ ***fines*** and in the ➔ ***coarse materials*** following a separation process.

Cut-Off

➔ ***Cut Size***

C-value

states by how many times the ➔ ***centrifugal acceleration*** in a ➔ ***centrifuge*** surpasses the gravitational acceleration g:

$$C = \frac{r\omega^2}{g} = \frac{n^2 d}{1800g}$$

r = centrifuge radius, ω = angular velocity, n = revolution number of the centrifugal drum, d = centrifuge diameter. C-values at industrial centrifuges can reach values of several 100 up to several 10,000.

Cyclone

➔ ***Hydrocyclone***

Dalton

Mass unit, named after the physicist Dalton, which is defined as a mass of a hypothetical atom with the atomic weight of 1. As mass unit the decimal multiples are applicable, for example instead of 1000 Dalton 1k-Dalton. At the ➔ ***ultrafiltration*** the ➔ ***cut-off*** is referred to the molecular mass instead to the particle size. This "molecular weight cut-off" (➔ ***MWCO***) is quoted in the unit "Dalton".

Damper

➔ ***Vibration Damper***

Darcy

One "Darcy" is the unit for measuring the permeability of a cube (side length 1cm) of a porous body at a pressure of 1bar and a liquid ➔ ***viscosity*** of 1mPa s. Dimension of the unit "Darcy" is (cm^2).

Darcy's Law

Fundamental equation for describing the laminar permeation of a porous layer with a Newtonian, incompressible fluid. The flow velocity v (➔ ***empty pipe velocity***) caused by the driving pressure difference Δp, through a ➔ ***bulk material*** with the thickness h_c and the specific ➔ ***cake resistance*** r_c with a liquid of the ➔ ***viscosity*** η_L amounts to:

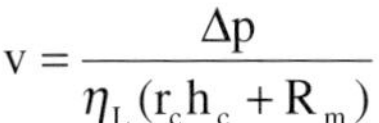

$$v = \frac{\Delta p}{\eta_L (r_c h_c + R_m)}$$

R_m denotes therewith the ➔ ***filter medium resistance***. If one applies Darcy's law to a compressible fluids, such as air or to a ➔ ***two phase flow***, the equation has to be adjusted according to the respective circum-stances.

Dead End Filtration

Discontinuous process variation of ➔ ***microfiltration*** or ➔ ***ultrafiltration***. With the dead end filtration the filtration process through a ➔ ***microporous membrane*** is continued until a given upper value of the pressure loss is reached, due to depositing of retained suspension contents. The filtration process has to be interrupted and the membrane has to be regenerated, e.g. through a periodic flow reversal. At very large volume flows, as in water treatment plants, the dead end filtration represents, in contrast to the ➔ ***cross-flow filtration*** with its necessarily large pump performances, the more feasible variant in terms of energy consumption.

Dead Flux

The flow of fine particles, smaller than the ➔ ***cut size***, discharged with the coarse underflow in a ➔ ***hydrocyclone***.

Debye Length

Term of the ➔ ***DLVO theory*** for the description of the characteristic penetration depth l_D of an electrical field, caused by surface charge on a suspended particle, into the surrounding ➔ ***electrolyte***:

$$l_D = \sqrt{\frac{\varepsilon kT}{2e^2 nz^2}}$$

ε = dielectric constant, k = Boltzmann-constant, T = absolute temperature, e = elementary electronic charge, n = density of the charge carrier, z = valence of the charge carriers.

Decanter

Continuously working solid bowl centrifuge in which a ➔ ***suspension*** is separated through ➔ ***sedimentation*** of the solids in the centrifugal field. The settled solids are removed out of the process room, which is conical tapered, by means of a ➔ ***conveyor screw*** at one end of the cylindrical drum. Hereby it turns with a differential number of revolutions compared with the main number of revolutions of the solid bowl drum. The exceeding clear liquid drains off over a usually adjustable weir at the opposite end of the solid bowl drum. Decanter centrifuges are offered in extremely different varieties, which conform according to the respective separation tasks. Thus decanter centrifuges exist for the pre-thickening of sludges (➔ ***Sedicanter***), for the maximum demoisturing of settled sludge (➔ ***Centripress***), for the separation of granular products, for ➔ ***classifying*** (➔ ***Centrisizer***) and for sorting (➔ ***Censor***). Furthermore decanters in a special design are also suited for the separation of three-phase mixtures (➔ ***Tricanter***), consisting of two liquids, incapable of mixing with each other (i.e. water and oil), and solid particles. Decanters are build with drum diameters of up to 2m and are operated with a number of revolutions of up to 10,000rpm. Large numbers of revolution per minute are applied with small rotor diameters for reasons of material strength. For the improvement of the sedimentation of difficult to separate substances the particles, to be separated in decanters, are often agglomerated as a preparative measure by means of ➔ ***polymeric flocculants***. Decanters are utilized in different industrial sectors and thus separate different products, such as granular PVC and compressible sewage sludge.

Decanter Centrifuge

➔ ***Decanter***

Deep Bed Filtration

Filtration process at which the particles to be separated accumulate inside a coarse-pored layer. Deep bed filtration should only be realized in the region of extremely low ➔ ***suspension concentrations*** to prevent an early blockage of the filter surface. They typically serve in the purification of minutely polluted liquids. Water and beverage filtration present large fields of application.

Deep Cone Thickener

Continuously working cylindrical or

conical gravitational thickener, characterized by construction heights of up to 30m, which are capable of producing high compression layers or compression pressures, and thus high underflow concentrations for thickened sludges.

Degerminating Filtration

➔ ***Sterile Filtration***

De-gritting

Pre-separation of coarse particles from a ➔ ***suspension***, which could upset the following separation process, done for example with a ➔ ***hydrocyclone*** or a ➔ ***sieve***.

Delayed Cake Filtration

➔ ***Crossflow Filtration***
➔ ***Dynamic Crossflow Filter***

Demixing

Non-homogeneous distribution of solids in a ➔ ***suspension*** or in a ➔ ***bulk***. Demixings can show up in terms of ➔ ***particle size***, particle shape and composition (density). The mass forces, such as gravity, are responsible for this, leading to different ➔ ***sedimentation velocities*** of the particles. Countermeasures against unwanted demixing can be a stirring apparatus or an increase of the ➔ ***suspension concentration***. High suspension concentrations lead, however, to ➔ ***swarm sedimentation***.

Demoisturing

Common expression in the solid-liquid separation for the removal of liquids out of a porous solids system capable off intra-particle force transmission due to close proximity. By comparison the removal of a liquid out of a ➔ ***suspension***, where the particles can still be moved against each other, is called ➔ ***thickening***. Solids systems, as presented by ➔ ***filter cakes*** or ➔ ***sediments***, can be demoistured either by gas displacement of the pore liquid or by the reduction of the void space through compression. The mechanical demoisturing is never complete as it ends at a ➔ ***mechanical demoisturing boundary***.

Demoisturing Angle

Expression from the field of ➔ ***rotary filters*** and here especially the ➔ ***drum-*** ➔ ***disc-*** and ➔ ***table filters***. The demoisturing angle α_2 describes the sector in which the ➔ ***filter cake*** that is formed in the ➔ ***suspension***, emerges out of the liquid and where it is exposed to a ➔ ***gas difference pressure***. In special cases the filter cake can also be demoistured by a ➔ ***press belt*** or ➔ ***belt rollers***. The demoisturing angle α_2 is connected over the filter number of revolutions n with the demoisturing time t_2 as follows:

$$t_2 = \frac{\alpha_2}{360^\circ}\frac{1}{n}$$

Demoisturing Equilibrium

State during the mechanical demoisturing of a porous solids system, estab-

lishing after the completion of the ➔ ***demoisturing kinetics***. At the equilibrium, the liquid retaining forces, such as the ➔ ***capillary forces***, compensate the liquid removing forces, for example the ➔ ***centrifugal forces***. Any further mechanical demoisturing can be achieved only by increasing the driving potential. All this is limited by the ➔ ***mechanical demoisturing boundary***.

De-fining

Removal of the fine grain fraction of a ➔ ***particle size distribution*** in a ➔***suspension***, for example with a ➔ ***hydrocyclone***. Purpose of the defining is either an improvement of the product's filtrability as it will be coarser then, or the liberation of the product from contaminants, which dominantly are in the fine grain region

Demoisturing Kinetics

Time dependent, degressive course of liquid removal from a porous solids system.

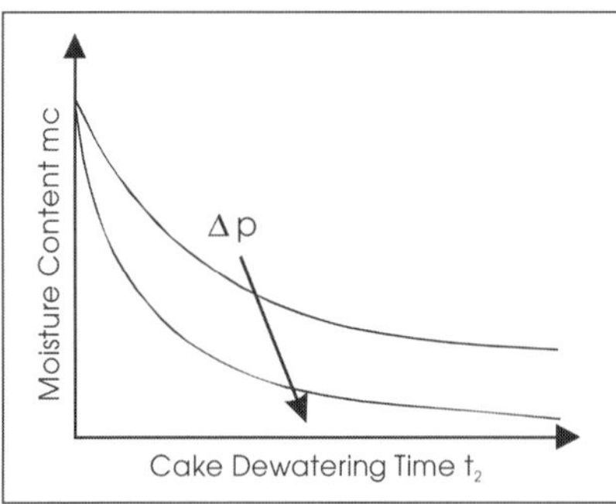

Filter cake moisture content as function of the dewatering time

A large quantity of liquid is removed at the onset, then the liquid outflow decreases due to the increasing influence of the liquid retaining forces, and finally it ends at the ➔ ***demoisturing equilibrium***.

Design Experiment

Testing on the laboratory or pilot scale for equipment scale-up. The design experiment has to simulate the separation process as planned in the full-size equipment, carried out with a representative product sample as realistic as possible.

Desorption

Detachment of substances (adsorbates) attached on a surface (adsorbent), by superceding the active forces (opposite of ➔ ***adsorption***).

Destabilization

A suspension in which particles exist individually and separate from each other is called a ➔ ***stable suspension***. Destabilization is understood to be a change of the electrical charges in a suspension in such a way, that the suspended particles can be agglomerated. Thus, the ➔ ***electrostatic repelling*** of the particles has to be reduced to an extent that the ➔ ***Van-der-Waals*** attraction, which is always present as a material-constant value, becomes dominant. This can be done for example by increasing the ion concen-tration or by changing the ➔ ***pH-value***. In water treatment, $FeCl_3$ is often added for this purpose.

Detergents

→ ***Tensides***

Diafiltration

Special processing technique in →***ultra-filtration*** where for example salts are separated from higher molecular solutions by adding new solvents commensurate with the flow rate of → ***permeate***, thus effecting a washing process.

Dialysis

is a membrane separation process. Low molecular side products (e.g. salts) are removed out of a solution of highly molecular substances (e.g. albumen, starch).

Diaphragm

Porous separator wall used for example in electrolysis between the cathode and anode cells. In the solid-liquid separation technology a diaphragm is understood to be a → ***membrane*** that is not only capable of retaining solid suspension components. For instance a → ***hydrophilic*** and → ***wetted*** → ***membrane*** can prevent the penetration of gas up to its → ***capillary entry pressure***.

Diatomaceous Earth

is employed as a → ***filtering aid***. As a natural product diatomaceous earth consists of finely granular skeleton fractions of diatoms and is characterized by a very large specific surface, therefore is capable of bonding turbid substances. It is extracted by open-pit mining in up to several 100m thick layers, processed and offered in → ***fractions*** of different granularity respective to the application. Due to its inert behavior diatomaceous earth is very often employed for the → ***filtration*** of beverages such as wine, beer, or fruit juices.

Differential Speed

Difference in revolutions generated with gears and drives by two co-axially linked running rotors. In several centrifuge types the solids are discharged from the process chamber with the aid of a → ***conveyer screw*** that is rotating with a differential speed in the direction of the centrifuge drum. Examples are the → ***decanter*** and the → ***worm screen centrifuge***.

Diffuse Double Layer

→ ***Electric double layer***

Diffusion

The spontaneously occurring mixing of substances, especially gases and liquids, that are directly in contact, and the equalization of their concentration differences in solutions. This is caused by the thermo-kinetic movement, whereby molecularly one substance penetrate into the other, i.e. they diffuse.

Dilatancy

→ ***Shear Thickening***

Dilution Washing

Process for the washing of a → ***filter cake*** by → ***re-mashing*** in → ***washing liquid***

and subsequent → ***filtration***. Dilution washing can be realized for example with in-series operating → ***drum filters***. By → ***re-suspending*** a → ***filter cake*** with a dry mass m_s, with a load w_1 ($w_1=m_L/m_s$) and a contaminant content X_1 (based on dry mass), in an amount m_W of washing liquid and subsequent filtration to a new load of w_2, the amount X_2 of contaminant in the new filter cake is:

$$X_2 = X_1 \frac{w_2}{w_1 + \frac{m_w}{m_s}}$$

Dipole Force

→ ***Hydrogen Bridge***

Directed Flow Screen Centrifuge

Continuous →***screen centrifuge*** with a conically widening rotor where the descending force component of the centrifugal forces acting on the particles convey them into the direction of the rotor outlet. Ring shaped assembled elements divide the glide path into guiding channels for better control as well as for lengthening of the path. A jagged edge on these channels continuously mixes the particles. Directed flow screen centrifuges are applied for → ***suspensions*** of higher → ***concentrations*** and coarser granularity (x<500µm).

Disc Filter

Continuously working → ***vacuum*** or → pressure filter for relatively easy to filter → ***suspensions*** and larger amount suspension flows. The radially sectioned filter discs are arranged on a horizontal filter shaft and often filter on both sides. Each of the sectional → ***filter cells*** is covered with a tightly stretched → ***filter bag*** and connected at the cell foot with a pipe to the → ***control head***, which controls the applied → ***vacuum***. The discs rotating at 0.5÷2rpm dip into a → ***filter trough*** filled with suspension, where the → ***filter cake*** is formed. As soon as the cake emerges from the →***suspension***, it is demoistured and dried with air. Finally, the cake is blown off by → ***pneumatic repulsion*** and discharged over a deflection plate. Alternatively, a basic → ***scraper discharge*** is used. Disc filters can have up to several 100m^2 filter area in a single machine.

Disc Package

Assembly of separating elements in → ***disc stack separators***. The circular-conical metal-sheet discs are stacked with bumper rods for a gap widths of approx. 1mm. In this manner the effective sedimentation area in the centrifuge can be increased significantly. For separating immiscible fluids the disc package is equipped additionally with vertical bore holes, which are called → ***rising channels***.

Disc stack separator

Continuously working sedimentation centrifuge where the available separation area is drastically increased through the installation of a → ***disc package*** into the centrifuge drum. In combination with extremely high → ***C-***

values up to 15,000, some disc stack separators can realize an ➔ ***equivalent clarifying area*** of several 100,000m^2. The discharge of the settled sludge through nozzles at the outer circumference of the dual-conical designed drum is either continuously (➔ ***nozzle separator***) or periodically (➔ ***self-cleaning separator***). Disc stack separators can be employed for solid-liquid, liquid-liquid, or solid-liquid-liquid separation, as well as for ➔ ***extraction***.

Discharge Chute

Simple duct or pipe, through which the separated solid leaves the separation equipment under the influence of gravity. For solid products that tend to adherence in the course of time a ➔ ***discharge screw*** is recom-mended because they can clog over time.

Discharge Screw

A transporting device, which removes by force a separated solid out of a separation apparatus. Discharge screws are employed especially when the solids are sticky, due to insufficient demoisturing, and a ➔ ***discharge chute*** would foul up.

Discontinuous Separation Apparatus

Employed for ➔ ***batch wise separation***, meaning a complete processing of individual volumes of ➔ ***suspension***.

Disperse

Distribution of the ➔ ***dispersed phase*** in the ➔ ***continuous phase***. The mixing of solids particles in a liquid for the production of a ➔ ***suspension*** is an example for this.

Disperse Phase

is a phase distributed as individual elements in another, contiguous surrounding phase (➔ ***continuous phase***. Examples for disperse phases are solids particles in ➔ ***suspensions***, gas bubbles in liquids, or liquid droplets in gas (mist).

Dispersion

A system, consisting of two or more phases, in which one phase (➔ ***disperse phase***) is evenly distributed in the dispersion medium (➔ ***continuous phase***).

Dispersivity

Term in the particle measuring technology, characterizing the physical property that is utilized for measuring. Thus the dispersivity characteristic can be a settling rate, a scattered light distribution, an attenuation, a voltage impulse, or similar phenomena. Through a physical relationship an ➔ ***equivalent diameter*** of particle can be determined out of such a dispersivity characteristic. An example would be the calculation of a particle diameter based on the settling velocity according to ➔ ***Stokes' law***.

Displacement Washing

denotes the cleaning of ➔ ***filter cakes*** by feeding a washing liquid. Under the influence of the driving potential, which can be either vacuum, gas over-

pressure, hydraulic pressure, or centrifugal pressure, the washing liquid is passed through the filter cake. The resulting removal of the ➔ ***mother liquor*** occurs in two different mechanisms. For one, the main amount of the ➔ ***pore liquid*** is displaced in plug-flow. The still remaining residuals of the substances to be removed are added to the flowing washing liquid by ➔ ***diffusion***, which is considerably more time intensive. Quality criteria for displacement washing are a high ➔ ***wash degree*** and a low consumption of washing liquid, which is expressed by the ➔ ***wash ratio***. An alternative process to the displacement washing is ➔ ***dilution washing***, where the filter cake is re-suspended in the washing liquid and subsequently filtered again.

DLVO-Theory

named after Derjaguin and Landau (Russia, 1941) as well as Verwey and Overbeek (Holland, 1948), is used for describing the stability of ➔ ***suspensions*** from a balance of the attracting ➔ ***Van-der-Waals forces*** and repelling ➔ ***electrostatic forces***. This theory is important for the solid-liquid separation technology as it allows to describe the state of suspensions, and agglomeration as well as ➔ ***flocculation*** of particles.

Double Acting Pusher Centrifuge

Special construction of a continuously working, single stage ➔ ***pusher centrifuge*** by the Escher Wyss company, at which the ➔ ***pusher plate*** is located in the middle of the sieve drum. The ➔ ***suspension*** is fed by means of a special device alternating to the front and backside of the pusher plate. The produced filter cake is transported by axially oscillating movements to the solids discharges at both drum ends. A cake ➔ ***washing*** with this design type is partially restricted. The special advantage of this machine is the large throughput capacity.

Double Belt Press

Continuously working ➔ ***press filter***, in which a ➔ ***filter cake*** is pressed out between two filter belts. A pressing power in the magnitude of 2÷4bar can be applied via press and deflection rollers, whereby the press belts are guided. Double belt presses can reach working lengths of several meters and are employed for the separation of difficult to filtrate and extremely compressible sludge. The feed slurries that are normally strongly flocculated have to be predemoistured by gravitational filtration in the ➔ ***straining zone*** so that they can be drawn between the belts. In the wastewater sector they compete mainly with the ➔ ***decanter*** and the ➔ ***filter press*** for sewage sludge separation.

Double Filter

Discontinuous, candle shaped ➔ ***sieve filter*** for the purification of liquids with minor amounts of particulate contamination. Double filters are mainly employed in the main flow with up to 100% filter area on stand-by. This can be by a put on flow by a three-way valving. An alternative to double filters present the ➔ ***automatic filters***.

Double-Flap Sluice Gates

Attachment employed in the solid-liquid separation for the discharge of demoistured solids from a continuous pressure filter system like the BOKELA → ***Hi-Bar***-Filter. Generally, the cylinder of double-flap sluice gates is filled first under the acting internal pressure of the filter apparatus with a closed outer gate; then the cylinder is isolated by the inner gate from the pressure chamber of the filter. It is emptied after the pressure has been released by opening the outer gate to atmospheric pressure. A special locking technique has to prevent that both gates opened simultaneously with the filter vessel still under pressure.

Double Layer

→ ***Electric double layer*** of → ***counter ions***, surrounding an electrically charged surface of suspended solid particles.

Double Weave

Weaves with different separation properties, strongly connected together. Double weaves display usually a fine-pored upper side facing the → ***suspension*** to ensure the retention of solids, and coarse underside lending stability to the media.

Drag Effect

observed especially with → ***decanter centrifuges***: particles are discharged with the clear liquid although they should have settled according to → ***Stokes' law***. Instead, the drag by the liquid overflowing the sediment causes a stirring up of already settled particles and carries them away.

Driving Potential

Force effect utilized for the separation of particles from of liquids, coming from → ***vacuum***, gas overpressure, or hydrostatic or centrifugal pressure, hydraulic, mechanical or → ***capillary pressure***.

Drum

Rotating, cylindrical element of a separation apparatus that is perforated and generally made from metals. In its interior or on its outer mantel area the solids are separated from the → ***suspension***. However, the drum used in sedimentation processes is impermeable and always fed in the interior.

Drum Filter

Continuously working rotary vacuum filter with a cylindrical, horizontal filter drum, often with → ***filter cells*** arranged coaxial on its mantel covered with a → ***filter cloth*** and connected through a filtrate pipe with the → ***control head***. The drum, rotating with approx. 0.2÷2 rpm, is immersed up to half of its diameter in a → ***filter trough***, where the feed → ***suspension*** is contained. The → ***filter cake*** forms on the filter cell while it moves through the suspension. When the cell emerges with the cake out of the suspension the cake is demoistured, provided that the → ***capillary pressure***, acting in the pores of the cake can be overcome by the applied → ***pressure difference***. Then the cake is removed from the drum either with a → ***scraper***,

by → ***compressed air repulsion***, or similar facilities. Drum filters are built with sizes up to 100m² → ***filter area***.

BOKELA ***drum filter*** with exchangeable filter cells (equipped with steam cabin for Hi-Bar Steam Pressure filtration)

Dry Substance

Mass related definition of the dry substance content in a → ***bulk*** following the separation process. The determination of the dry substance DS is simple and is performed by weighing, respectively, the moist and the dry → ***cakes***. The mass of the solids m_S is then related to the total mass of the moist → ***bulk*** m_{tot}, made up of the solids mass m_S and the liquids mass m_L. The dry substance is quoted in weight percentage. Substances of different densities cannot be compared on the basis of their dry substance values.

$$DS = \frac{m_s}{m_s + m_L}$$

Dynamic Buoyancy

→ ***Buoyancy***

Dynamic Crossflow Filter

→ ***Crossflow-filter*** with a shear flow, adjustable independently from the pump-pressure. The flow is generated by a rotor-stator system, where the suspension is processed in its annulus. Rotor as well as stator or both can be designed for filtering. There are axial and coaxial dynamic crossflow filters. The → ***DYNO-Filter*** of the BOKELA company represents a special form of the radial dynamic crossflow filters.

Dynamic Filtration

Alternate expression for the → ***crossflow filtration***, whereby the feed suspension flows tangentially to the → ***filter medium***. A special design in dynamic filtration is the → ***DYNO-Filter*** by the BOKELA company.

Dynamic Sieve Filtration

Innovative process of → ***sieve filtration*** by the BOKELA company based on the principle of → ***dynamic filtration*** with the → ***DYNO-Filter***. At dynamic sieve filtration the DYNO-Filter is used for a continuous separation of coarse particles from suspensions. The fine → ***particle fraction*** and the → ***mother liquor*** pass the sieve medium while the coarse fraction is held back by the sieve and discharged as highly concentrated

residual suspension.

Dynamic Viscosity

refers to the viscousness or internal friction of a fluid. It characterizes a material property, according to which tangential forces appear, acting against a parallel displacement of liquid or gas layers relative to each other. The dynamic viscosity is measured in (N/m^2) or (Pa). It is the force required to flow for a thin layer of gas or liquid of unit size area with a velocity that is 1m/sec higher than that of a layer 1m away. Viscous substances have high viscosities, and low-viscous ones have lower viscosities. The dynamic viscosity of water is η=0,001kg/ms.
$\eta(Pa{\cdot}s = N{\cdot}s/m^2 = kg/m{\cdot}s = 10^3cP)$

DYNO-Filter

→ ***Dynamic filtration*** machine by the BOKELA company featuring up to 12 m^2 filter area for the dynamic → ***crossflow filtration*** across discs. It is employed for → ***thickening***, → ***classifying*** (→ ***sieve filtration***) and → ***washing*** of suspended particles. The modular structured machine comprises a sequence of disc-shaped filter chambers, with stirring elements inside on a common central shaft that rotates to prevent cake formation.

DYNO-Filter with 12m^2 filter area

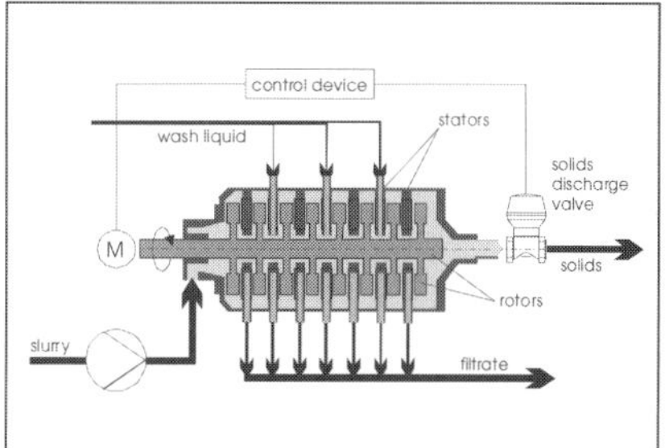

Schematic sketch of the ***DYNO-Filter***

Electric Double Layer

An electric double layer around suspended particles exists due to the fact that the particle surfaces carry an electric charge. To this surface adheres a respectively opposing, loaded ion layer (➔ ***Stern's layer***), compensating the particle charge extensively but not totally. Complete neutralization is achieved through a diffuse, further liquid layer, displaying a slight excess of the respective counter-ions. Only outside of this "double layer" are the charges in the ➔ ***suspension*** balanced again.

Electro-coagulation

Process for the ➔ ***agglomeration*** of fine-grained particles that carry a surface charge in liquids and therefore form at first a stable ➔ ***suspension***. By application of an electrical field the particles are moved respective to their charge to the opposing electrode, where they are then discharged and subsequently are capable of agglomerating due to ➔ ***Van-der-Waals*** forces. The process is especially of interest when additional substances, such as ➔ ***flocculation agents***, are disallowed in the system.

Electrofiltration

includes as a collective term various electro-kinetic effects that are produced when a ➔ ***suspension*** is exposed to an electrical field. If charged solids particles move in a suspension, then it is called ➔ ***electrophoresis***. The movement of liquid inside of a solids structure under the influence of an electrical field is called ➔ ***electro-osmosis***.

Electrolysis

is the decomposing of electrically conductive liquids (➔ ***electrolyte***) by current flow into their constituents.

Electrolyte

Collective term for liquids, which are conductive due to molecules dissociated into ions.

Electroosmosis

Electro-kinetic phenomenon by which water can be removed from a porous solids system, if this is in contact with an electrical field. A mechanism for the liquid transport derives from the fact that the ions contained in the liquid, can form hydrate shells and in this manner carry water to the electrode during their migration. Electroosmotic effects have been or are currently applied in the desalinization of water, demoisturing of peat, pigments, or colloids (e.g. latex, glue), for the cleaning of clay, silicic acid, or for the desiccation of moist brick walls.

Electrophoresis

Process, at which molecular disperse or ➔ ***colloid disperse*** particles of a solution migrate in an electrical field, i.e. respective to their charge to the cathode (cataphoresis) or to the anode (anaphoresis). Subsequently, they can be separated from each other.

Electrostatic Repelling

If suspended solid particles carry an electrical charge then the particles, often charged with the same polarity, will repel each other. Therewith the electrical potential shields off so effectively the ➔ ***Van-der-Waals forces*** originating from the particles, that the moving particles are prevented from adhering to each other and the ➔ ***suspension*** remains stable.

Emergency-Off

Safety switch installed in the working area of an operator for activating an immediate stand still (➔ ***shut down***) of the apparatus in the case of emergency.

Empty Pipe Velocity

Flow velocity of a fluid, establishing before and behind a ➔ ***bulk***, if one assumes the cross-sectional area as being perpendicular to the flow direction. The mean flow velocity in the ➔ ***bulk*** is enlarged by the contraction of the cross sectional area and can be found by dividing the empty pipe velocity by the bulk porosity.

Emulsion

➔ ***Dispersion*** made of two immiscible liquids, where one is distributed in form of small, stable drops in the other liquid.

Endurance

➔ ***service life***

Entry Cross Section

Area in the feeding device of a separation apparatus that is perpendicular to the flow direction of the suspension.

Equilibrium Load

➔ ***Equilibrium Moisture***

Equilibrium Moisture

An equilibrium moisture of a ➔ ***filter cake*** or of a ➔ ***sediment*** is established when, following the completion of the ➔ ***demoisturing kinetics***, the driving potential such as a gas difference or a centrifugal pressure is in equilibrium with the opposing potential, i.e. the ➔ ***capillary pressure*** or the solids pressure. The equilibrium moisture represents for a given demoisturing potential the physically minimally reachable moisture of a product. Respectively to the definition of the moisture degree in the ➔ ***bulk*** one talks either about equilibrium residual moisture, equilibrium saturation, or equilibrium load. In the course of a technical demoisturing process there is generally not enough time available to reach this equilibrium. The equilibrium moisture, however, that is determined in the laboratory indicates the physically

possible demoisturing potential at a certain pressure.

Equilibrium Residual Moisture

➔ ***Equilibrium Moisture***

Equilibrium Saturation

➔ ***Equilibrium Moisture***

Equivalent Cake Thickness

Term originating from ➔ ***cake filtration***. The equivalent cake height h_{ce} characterizes the applied ➔ ***filter medium*** with a cake layer of the same flow resistance. It is calculated, respectively, as the product of the specific ➔ ***cake permeability*** p_c and the ➔ ***filter medium*** resistance R_m, or as the quotient of filter medium resistance R_m and specific ➔ ***cake resistance*** r_c:

$$h_{ce} = p_c R_m = \frac{R_m}{r_c}$$

Equivalent Clarifying Area

Term applied in ➔ ***sedimentation centrifuge*** technology. The equivalent clarifying area Σ of a sedimentation centrifuge indicates how many m^2 clarifying area A in the earth's gravitational field can be substituted by it, if one applies ➔ ***Stokes' law*** for expressing the ➔ ***settling velocity*** of the particles:

$$\Sigma = AC$$

C is herein the ➔ ***C-value***, indicating the multiple of the earth's acceleration g, which can be realized inside the centrifuge in question.

Equivalent Diameter

The equivalent diameter of a particle or a pore is a measure of a particle or a ➔ ***pore*** with defined geometrical shape (e.g. circle) with the same characteristic property (e.g. area) as the observed characteristic of the investigated particle or pore. The settling velocity-equivalent diameter of a particle of any shape for example corresponds to the diameter of a sphere with the same ➔ ***settling velocity***.

Excess Pressure

➔ ***Overpressure***

Extract

In ➔ ***extraction*** selectively enriched component from a ➔ ***suspension*** in a dissolution process with an ➔ ***extraction agent*** that is not miscible with the suspension liquid.

Extraction

Process for the complete or partial separation of a liquid or solid substance mixture by means of a solvent or ➔ ***extraction agent*** that is not miscible with the suspension liquid. The components of the substance mixture to be removed must have different solubilities in the solvent and the ➔ ***extraction agent***. The extraction is a selective process, i.e. the extraction agent has the capability to accept only certain preferred substances.

Extraction Agent

A liquid capable of incorporating certain substances selectively by solution during ➔ ***extraction***.

Feed Cross Section

Opening area of the feed device for the ➔ ***suspension*** to be separated, in a separation apparatus.

Feed Material

Generalizing term for ➔ ***suspensions*** fed into process equipment.

Felt

➔ ***Needled Felt***

Film Flow Model

The film flow model was conceptualized for the centrifugal demoisturing of a ➔ ***bulk***, especially those with a coarser granularity. This model distinguishes between a ➔ ***plug flow*** mode and a subsequent film flow of the liquid remaining on the particle surfaces. It was inspired by the image of a plate, that when pulled out of an oil bath shows an oil film varying locally and with time in thickness until an even flow is established.

Filter Area

is the active area in a filtration process, which is covered with a ➔ ***filter cloth***. In a ➔ ***drum filter***, for example, the entire surface area covered with filter cloth is active as filter area. In a ➔ ***vacuum belt filter*** with a ➔ ***rubber conveyor belt*** it includes only that part of the filter cloth that is on the upper side of the filter, whereas the other half is out of use as it is led back on the underside of the belt.

Filter Bag

➔ ***Filter medium*** tailored as a bag, as employed in ➔ ***bag filters*** or ➔ ***disc filters***. Depending on the design of a filter apparatus, filter bags are flown through from the inside to the outside or from the outside to the inside.

Filter Cake

A porous layer of solids particles formed on the surface of a ➔ ***filter medium***. In order to generate a filter cake, ➔ ***solids bridges*** have to develop over the media openings thus blocking the following solids. Filter cakes are formed in ➔ ***vacuum filters***, ➔ ***pressure filters***, ➔ ***press filters***, and ➔ ***filter centrifuges***. This can occur discontinuously or continuously, and under a constant or changing driving potential. Filter cakes are subjected in general to an extensive mechanical ➔ ***demoisturing*** after ➔ ***cake formation***, and as the case may be are liberated of still remanent ➔ ***mother liquor*** by ➔ ***washing***.

Filter Candle

Cylinder-shaped filter element of a ➔ ***candle filter***, mostly consisting of a

perforated internal pipe and a filter medium placed on top. As a rule filter candles are permeated from the outside to the inside. Respective to the type of filter medium filter candles can be employed for ➔ ***cake filtration*** or ➔ ***deep bed filtration***. The ➔ ***filtration*** on a filter candle normally is a discontinuous process.

Filter Cartridge

Changeable, finite filter element, preferably employed for ➔ ***deep bed filtration*** and/or membrane filtration that is replaced after reaching a critical contamination.

Filter Cell

The filter surface in continuously working ➔ ***rotary filters*** is divided into single cells, each separately connected to the ➔ ***filtrate pipe system***. The filter cells in ➔ ***drum filters*** are flat, rectangular pans with a plastic ➔ ***cell inlay*** on which the ➔ ***filter cloth*** is attached, often with a ➔ ***back cloth*** between. In ➔ ***disc filters*** the filter cells are shaped as circular segments and designed to filter on both sides. Usually, they have a perforated metal over which the filter cloth is pulled as a tailored bag and attached. The ➔ ***filtrate*** is drawn off at the narrow end of the cell, which is called the cell foot.

Filter Centrifuge

➔ ***Centrifugal Filter***

Filter Cloth

Special type of ➔ ***filter medium***, consisting of weave. There is a plethora of different materials, different thread or respectively yarn quality as well as extremely different weave structures. The filter cloth represents the critical interface between separation equipment and suspension. It must be selected carefully for each individual application to meet the process, mechanical, and chemical demands, respectively. The selection of a filter cloth should never occur solely by theory, but instead has to be supported experimentally.

Filter Cloth Blockage

➔ ***Filter Cloth Clogging***

Filter Cloth Clogging

A media-fouling mechanism in ➔ ***cake filtration***, where over the course of operation more and more solids penetrate the structure of the weave and deposit in it. Such cloggings are some times partially reversible and can be undone by ➔ ***filter regeneration***, or if they are irreversible will render the filter cloth unusable, as evidenced by a loss of ➔ ***permeability***.

Filter Cloth Resistance

Flow resistance R_m of a ➔ ***filter cloth***. It can be determined directly by ➔ ***Darcy's law***, if one permeates the filter cloth with a particle-free liquid at a specified pressure and measures the flow. However, this value is invalid for predicting the filtration performance with a ➔ ***suspension***, because the interaction proper between filter cloth and suspension with particles settling in in the cloth structure and especially with

a ➔ ***bridge layer*** forming leads to a relevant filter cloth resistance. Thus, it can only be measured in conjunction with the actual suspension and be estimated for example by the ➔ ***t/V= f(V)-Method***.

Filter Fineness

Data about the separation efficiency of a ➔ ***filter medium*** at ➔ ***filtration***. The nominal fineness notes the particle percentage of a certain size that is retained by the medium. However, test materials and experimental conditions influence strongly this value so that nominal filter fineness data from different filter manufacturers are difficult to compare. Information about the test material and experimental conditions has to be included when stating a nominal filter fineness.

Filter Medium

Porous layer, through which during the ➔ ***filtration*** the ➔ ***filtrate*** permeates due to an acting ➔ ***pressure difference***, while the solids in the original suspension are retained on its surface or in its structure, respectively. Filter media can be of very different types: metallic sieves, textile fabrics, fiber ➔ ***fleeces***, felts, paper, membranes, sintered materials, ➔ ***bulk*** layers, and a host of others are employed. An optimal filter media has maximum separation efficiency, minimal flow resistance, and a long service life. In actual operation a supportable compromise has to be found among these contradicting objectives for each individual separation task.

Filter Medium Resistance

➔ ***Filter Cloth Resistance***

Filter Optimization

➔ ***Revamping***

Filter Paper

A ➔ ***filter medium***, that is preferred in the laboratory for preparative purposes. On the other hand, a technical filter medium is recommended for industrial filter tests, since the filter medium can influence the filtration process significantly.

Filter Performance

➔ ***Throughput***

Filter Press Membrane

Plastic or rubber membranes, encased in a special membrane plate and applied in ➔ ***membrane filter presses***. The membrane is pressed with a pressing liquid, which is pumped into the empty space between membrane and plate wall, against the ➔ ***filter cake*** and squeezes it mechanically.

Filter Press

One of the most commonly used solid-liquid separation machines for the discontinuous ➔ ***cake filtration*** and demoisturing of difficult to filter ➔ ***suspensions***. A filter press has a frame with a stack of filter plates packed between head and end piece, which are pressed together for sealing. In the case of a ➔ ***chamber filter presses*** rectangular or

square plates with a side length of up to 2m, form filter chambers, lined with a ➔ ***filter medium***, into which the feed suspension is pumped. In a ➔ ***frame filter presses*** the filter chambers are created by alternating special frames alternating with the filter plates in the plate package. The cake of ➔ ***membrane filter presses*** can be squeezed mechanically by a rubber membrane on one side. A filter cycle can take anywhere from of a few minutes up to several hours. Filter presses have areas of up to 1,000m^2 and operate generally with pressures of up to 16 bar. High-pressure filter presses are applied with up to 60bar. Due to their almost universally applicability for difficult to filtrate suspensions filter presses have an extremely large field of operation.

Filter Reactor

Discontinuously working ➔ ***pressure filter*** capable of performing, next to the pure separation task, also such unit operations as reaction, crystallization, extraction, thermal drying et al. Generally, an agitator is installed in filter reactors. Some filter reactor types can be also rotated or tilted for the beneficial execution of a specific processing step. Filter reactors offer great advantage if a contamination of the products or an exposure to the environment has to be avoided, because all the different steps take place in a minimized and also well-isolated space. An example for a filter reactor is the Titus-Filter-Dryer by the KRAUSS MAFFEI company.

Filter Regeneration

Term used for all types of ➔ ***filtration*** to describe the restoration of a ➔ ***filter medium's*** ability to perform after getting clogged with particle deposits or crystallized products. Back flushing, application of ultra sound, blasting with high-pressure water, chemical cleaning with acids, or washing with hot water are some of the measures applied. In extreme cases filter regeneration can mean replacing the filter medium, i.e. when it cannot be cleaned.

Filter Trough

Tub-shaped container filled with a suspension wherein the filter cells of ➔ ***drum filters*** or ➔ ***disc filters*** are periodically immersed in order to form a ➔ ***filter cake***. The filter trough generally has either a pendulum-type agitator, a propeller, or a paddle agitator to homogenize the ➔ ***suspension*** and to keep the particles in suspension. In disc filters extremely narrow troughs are known, where a mechanical stirring apparatus can be omitted, as the rotational movement of the discs often provides enough mixing of the suspension.

Filter Segment

expression for the ➔ ***filter cells*** of a ➔ ***disc filter***

Filterability

Evaluation criterion for separating of a suspended solid by ➔ ***filtration***. Often employed for this is the specific ➔ ***filter cake resistance*** as determined by a

filtration test with the ➔ ***t/V-over-V method***.

Filter Aids

Additives to enhance the filterability of ➔ ***suspensions***. Filter aids can be applied as a ➔ ***precoat layer*** on the filter medium prior to the actual filtration, or as ➔ ***body feed filtration*** admix to the suspension to be separated. Common filter aid materials are: ➔ ***diatomaceous earth***, ➔ ***perlite***, wood flour, ➔ ***activated carbon***, cellulose fibers and others. They are beneficial for suspensions that form extremely difficult to permeate ➔ ***filter cakes***. Major application areas are water treatment and beverage purification.

Filtrate Collector

➔ ***Receiver***

Filtrate Pipe

Pipe for the discharge of a liquid separated by ➔ ***filtration*** out of an ➔ ***suspension***.

Filtrate Pipe System

for transferring the ➔ ***filtrate*** out of a filter apparatus. In a more specified usage it refers to the filtrate pipes in ➔ ***rotary filters***, that exist in various manifolded designs for connecting individual ➔ ***filter cells*** with a single ➔ ***control head*** of the filter. When sizing the filtrate pipe system the minimization of its pressure loss has priority, as this share of the acting ➔ ***pressure difference*** is excluded from the actual filtration process. The performance of many filters, however, is limited by an undersized filtrate pipe system dimensioned too narrowly.

Filtrate

The liquid, separated by a filtration process, previously having permeated the ➔ ***filter medium***.

Filtratest

Portable laboratory pressure filter apparatus of the BOKELA company for investigating vacuum, overpressure, and press type filtrations. It is based on the principle of the discontinuously working ➔ ***monoplate filter***, has a filter area of 20 cm^2 and can be pressurized up to 11 bar. All process steps of the ➔ ***cake filtration*** like the cake formation, cake washing, cake demoisturing can be investigated. An electronic measurement data registrating and processing system records the filtration data and issues them in the form of an experimental test protocol.

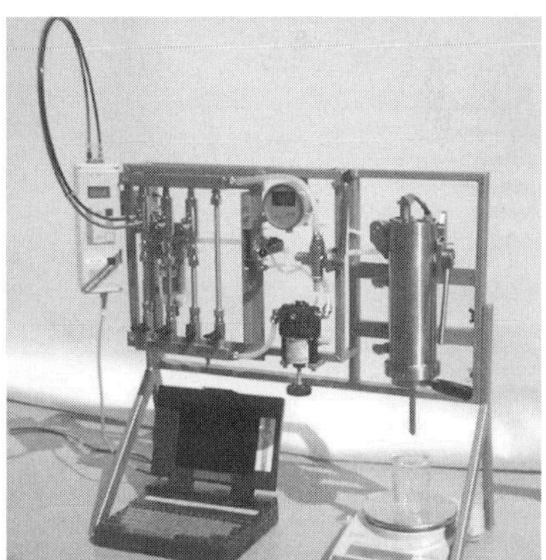

Filtratest

The apparatus corresponds to the VDI-guideline No. 2762. Furthermore, it can be employed for measurements according to DIN 38409, ASTM F317-72 and ASTM F 1170-88.

Filtration

In the field of solid-liquid separation filtration is defined as the segregation of a ➔ ***suspension*** into solids and ➔ ***filtrate volumes*** by means of a porous ➔ ***filter medium***. Both solids and liquid are moved by a driving pressure gradient in co-current flow towards the filter medium. The liquid is able to penetrate the filter medium, while the solids are retained. Filtration is next to the ➔ ***sedimentation*** one of the basic principles applied in solid-liquid separation. According to the mode of execution, one distinguishes further between ➔ ***cake***, ➔ ***crossflow***, and ➔ ***deep bed filtration*** modes.

Filtration Pressure

is the active driving pressure gradient during a ➔ ***filtration***. The filtration pressure can be generated by a hydrostatic head, centrifugal forces, applying a vacuum, charging a gas overpressure, the head of a slurry pump, or the pressure of an impermeable press membrane. The filtration pressure is either kept constant during filtration, or is increased proportional to the pressure loss increase caused by the cake formation.

Fine Capillary System

In a ➔ ***bulk*** one typically differentiates between a coarse and a fine capillary system. It should be pointed out that he term "fine capillary system" can mean different things, i.e. the pore structure of ➔ ***agglomerates***, the pores in a solids particle, or the ➔ ***interstitial liquid*** remaining in the ➔ ***bulk*** after a mechanical demoisturing process.

Fine Filter

Class of filters belonging to ➔ ***deep bed filters*** characterized by solid, porous bodies made of sintered metal, plastic, or ceramic powder. They typically have a mean pore size far below 1 mm, so that they are suitable for deep bed filtration of extremely fine particles down to the µm region. ➔ ***Leaf filters*** and ➔ ***candle filters*** with ➔ ***precoat layer*** are also employed as fine filters. A distinguishing performance objective of fine filters is a ➔ ***filtrate*** as particle-free as possible.

Fines

Particle collective formed during the separation of a ➔ ***feed material*** with a certain ➔ ***particle size distribution***: fines entail the fraction with the smaller size, and the ➔ ***coarse material*** con-taining the larger particles. The particle size at the boundary between fines and coarse material is called the ➔ ***cut size***.

Fingering

is a phenomenon in ➔ ***cake filtration*** often occurring during the ➔ ***washing*** or the ➔ ***demoisturing***. The washing liquid, or the gas, respectively, does not progress in a uniformly even front through the cake, but preferentially penetrates the larger ➔ ***pores***, that are always present in normal particle size

distributions due to their lower resistance. This can lead to fingering, i.e. over parts of the filter area there is a premature and unwanted break through of the washing liquid (or gas) through the ➔ ***filter medium***. Counter measures are limited: an equalization of the cake structure and secondly as high as possible a viscosity of the washing fluid compared to the ➔ ***mother liquor*** can reduce the fingering effect. ➔ ***Steam pressure filtration*** has shown to alleviate the fingering when through condensation of the pressurized steam the liquid in the advancing, large pores can be replenished.

Fixing Wire

serves as an additional attachment of the ➔ ***filter cloth*** on ➔ ***drum filters***, wrapped spiral-like around the cylinder, specially when a ➔ ***compressed air repulsion*** is utilized.

Flake

➔ ***Flock***

Flat Bottom Cyclone

Cylindrically shaped ➔ ***hydrocyclone*** with a flat, non-conical bottom wherein an induced convective flow carries the solids into the center of the flat bottom, from where they can be removed as a thickened ➔ ***sludge***. By controlling the sludge height in the cyclone the desired ➔ ***cut size*** can be freely adjusted within certain limits. Flat bottom cyclones can classify solids up to a cut size of 500µm.

Fleece

➔ ***Filter medium*** made of compressed fiber layers. A fleece compacted by needling is also called ➔ ***needle felt***. Fleeces are low cost media, with relatively low strength, however, and lack a regular pore shape or precisely adjustable pore size in contrast to a ➔ ***weave***.

Flocculation

Process for the aggregation of ➔ ***primary particles*** to more or less loosely constructed particle collectives named ➔ ***flocks***. Flocculation improves the separation behavior of finely granular ➔ ***suspension*** through an increase of the sedimentation velocity of the solids or by forming more permeable ➔ ***filter cakes***. For the flocculation of a suspension, a parent solution has to be produced at first from a ➔ ***polymeric flocculant***, diluted to the operational concentration, and is then added to the suspension to be flocculated. A rapid and thorough admixing is required to bring the polymers uniformly into contact with all particles. A flocculation holding period follows, in which the actual joining of primary particles takes place under low stress.

Flocculation Agents

Chemicals, which upon addition to a ➔ ***suspension*** lead to the aggregation of individual particles to larger particle collectives, also called ➔ ***flocks***. Flocculation agents mostly are long-chained polymers which carry electric charges which are either neutral (nonionic), anionic, or cationic with

respect to the solids to be flocculated. Polymeric flocculants are supplied in form of powders or concentrated solutions. Generally, they can't be applied in the food sector. Flocculation agents are a significant operating cost factor in wastewater treatment. They are applied in particular with ➔ ***thickeners***, ➔ ***decanter centrifuges***, ➔ ***sieve belt presses***, and ➔ ***filter presses***. Many applications are also found in the area of continuous vacuum filtration.

Flocculation Aids

➔ ***Flocculation Agents***

Flocculation Holding Period

➔ ***Flocculation***

Flock

is an aggregate collective of ➔ ***primary particles***. The ➔ ***adhesion*** of the particles in a flock is effected by either influencing the electric surrounding of the particles (➔ ***agglomeration***), or by adding ➔ ***polymeric flocculating agents*** (➔ ***flocculation***). A flock does sediment faster than the single particles would, ties in the finest, suspended matter, and produces during ➔ ***cake filtration*** a cake structure with a higher permeability. ➔ ***Sediments***, produced out of flocks or ➔ ***filter cakes*** are considerably more compressible than structures, formed out of comparable single particles

Flock Density

Packing density of ➔ ***primary particles*** in a ➔ ***flock***. The more compacted a flock is, the higher is its strength against mechanical stress. For this reason the flock density of a flocculated suspension is increased prior to feeding into a ➔ ***centrifuge*** by a shearing pretreatment, e.g. in a cylindrical stirrer.

Flock Factor

used in the ➔ ***Richardson & Zaki-equation*** to determine the volume fraction of ➔ ***flocks*** in a flocculated ➔ ***suspension***. The volumetric flock concentration c_{vF} is derived as the product of ➔ ***solids volume concentration*** c_v and flock factor k:

$$c_{vF} = kc_v$$

Flow Moisture Point

Test method for estimating the stability of a wet ➔ ***bulk material*** under an alternating load. A semi-spherically shaped test sample of moist material is exposed to defined vibrations on an oscillating table. At some moisture content the ➔ ***bulk*** starts to deform and begins to flow. This test method serves the purpose to assure stable storage conditions during the transport of moist ➔ ***bulk materials*** by truck, train, or ship.

Fluid

General term for a liquid or a gas.

Forced Discharge

through a special discharge device for demoistured solids to guaranty the safe removal of the product under any circumstance, e.g. a ➔ ***discharge screw***.

Forward Edge

The forward edge of ➔ ***filter cells*** at ➔ ***rotary filters*** in the direction of the rotation, characterized by the fact that it has the shortest cake formation time, i.e. the ➔ ***cake thickness*** there is at its lowest value. If the filter cells are not too wide this effect is not significant. At ➔ ***disc filters*** with relatively few cells, however, the zone of the forward edge, that is close to the cell foot, can cause a problem during the demoisturing as more than 50% of the total amount of air can penetrate this extremely small filter area.

Fouling

Dirt layer, formed by biological activity, on ➔ ***membranes***. Fouling can lead to a clogging of the membrane pores and therewith to a decrease of the ➔ ***permeate flow***. Fouling can be abated by different chemical, physical and mechanical regeneration measures.

Fraction

Clearly differentiated subset in a particle collective defined by their ➔ ***particle size*** or type of particle.

Fractional Grade Efficiency

Term for describing the separation characteristic T(x) of a separation apparatus also called Tromp curve. It represents the particle amount $M_G(x)$ with the defined particle size x as percentage of the total amount $M_A(x)$ of a particle collective, which is discharged in the ➔ ***coarse material*** of the separating apparatus.

$$T(x) = \frac{M_G(x)}{M_A(x)}$$

Frame Filter Press

Oldest design of a ➔ ***filter press*** characterized by the feature, that the chamber for holding the ➔ ***filter cake*** is formed by a frame, which constitutes the plate package, together with alternating filter plates between frames. A disadvantage of the frame filter press is the solids discharge, which has to be manually performed by breaking the cake out of the frame. An automation of this step was initially introduced with the ➔ ***chamber filter press***.

Friction

Force of resistance, counteracting the movement of one body along the surface of another one. According to Coulomb's law, the frictional force F depends on the coefficient of friction μ and the normal force N, with which the surfaces in contact press against each other. However, it is not dependent on the size of these surfaces:

$$F = \mu N$$

As long as there is no movement, i.e. at adhesive friction, μ is larger than after the onset of sliding, i.e. during sliding friction.

Fundabac Filter

Special design of a discontinuously working ➔ ***cake filter*** by the DrM, Dr. Müller company. The ➔ ***filter candles***

consist of six perforated tubes surrounding a central dip tube. The sock shaped filter media attached to both ends of the candle retains the solids on the outside while the filtrate is guided to the bottom of the outer tubes and upward through the dip tube. The candles are manifolded to registers in a pressure tank with typical differential pressures of 4 – 5 bars. Fundabac filters are suited for wet as well as dry discharge of a product and can also be employed as ➔ ***precoat filters***, with in-situ cake blow back and sock cleaning by automatic sequencing. The design without a tubesheet offers in addition heel filtration and cake washing, and the Contibac design variant is for quasi-continuous processing. Both filter types, offered with up to 1,700 sq.ft. filter area, are representatives of ➔ ***fine filters*** for dilute, difficult to filter ➔***suspensions***.

Gas Throughput

Term from ➔ ***cake filtration***. During the demoisturing process under the influence of the gas difference pressure and following the demoisturing of the largest ➔ ***pores*** a gas throughput through the respectively demoistured pores of the ➔ ***filter cake*** occurs. With decreasing product moisture the gas throughput increases. It has to be sustained by the gas compressor, in order to maintain the driving pressure for further cake demoisturing. The gas throughput is not the cause but instead the undesired effect of the demoisturing. The gas throughput determines essentially the operational costs of the filter apparatus.

Gas Throughput-Free Filtration

Process developed in Karlsruhe, Germany for ➔ ***cake filtration*** based on a ➔ ***membrane*** ➔ ***filter cloth*** with a high ➔ ***capillary entry pressure*** as ➔ ***filter medium*** (BOKELA patent). The process avoids through the structure of this ➔ ***filter medium*** the ➔ ***gas throughput***, normally occurring at the cake demoisturing through the already emptied cake pores. Furthermore particle-free ➔ ***filtrates*** are produced. The filtration free-of-gas throughput can be employed for all cake filtration machines due the flexibility and process-tailored fabrication of the filter medium.

Grade Efficiency

Share of percentage of a substance having been separated following the separation. The ➔ ***total degree of separation*** for a solid substance out of the ➔ ***suspension*** includes the separated amount in percentage share of the entirely available solids, while the ➔ ***fractional grade efficiency*** describes the percentage of the separated share of a solids fraction. The fractional grade efficiency is displayed by the ➔ ***Tromp curve***.

Grain

➔ ***Particle***

Grain Fraction

➔ ***Particle Fraction***

Grain Size Distribution

➔ ***Particle Size Distribution***

Grain Size

➔ ***Particle Size***

Gravity

The gravity measured on the earth's surface is derived as the resultant out of the mass attraction and the ➔ ***centrifugal force*** caused by the earth's rotation, which in general can be neglected.

Gravity Filtration

The hydrostatic pressure of a liquid column is utilized here as the driving potential for the liquid transport. This can be realized by an over-the-dam height on the surface of a → ***deep bed filter*** (→ ***sand filter*** for water purification), or by the liquid column in a → ***bulk***

Gravity Thickener

Mostly round tank in which a feed → ***suspension*** is separated by → ***gravity***. Round thickeners are built with diameters of up to 200m and are operated continuously. The diluted, often flocculated, suspension is fed centrally. Below the clear liquid zone in the upper thickener section whose outer edge includes the overflow for the clarified liquid, is the interface level where the so-called → ***swarm sedimen-tation zone*** begins. This separating zone changes into the → ***compression zone*** in the lower part of the tank. Here the particles approach each other so closely that they are capable of exerting mechanical forces on each other. The achieved thickening degree depends on the thickness of the compression layer and the compression time. The thickened sludge is conveyed from the bottom of the thickener by a slowly rotating → ***rake*** to the central sludge outlet. Very high thickening is achieved in so-called → ***deep cone thickeners***.

Gupte-Equation

Permeation equation for porous → ***bulk*** with a similar structure as the → ***Carman & Kozeny-equation***:

$$\frac{\Delta p}{\rho \bar{v}^2}\frac{d}{h_c} = \frac{5.6}{Re}\varepsilon^{5.5} \quad ; Re < 1$$

Δp = pressure difference, ρ = fluid density, v = mean (average)flow velocity, d = characteristic length, h_c = cake thickness, ε = porosity, Re → ***Reynolds Number***

HBF

➔ *Hyperbar Filter*

Helmholtz Vortex

➔ *Potential Vortex*

Hi-Bar Filtration

Development by the BOKELA company. Continuously working pressure filter system according to the principle of the ➔ ***Hyperbar filter***. The Hi-Bar-Filter can also be designed as a ➔ ***steam pressure filtration.*** A special variant of Hi-Bar fiiltration is the ➔ ***Oyster Filter.***

BOKELA ***Hi-Bar*** Filtration pilot plant with 2 shipping containers

High Intensity Press

Post-demoisturing apparatus, by the ANDRITZ company, for mechanically pre-demoistured, compressible materials. The high intensity press works completely continuously on the principle of the ➔ ***double belt press***, however without the common multiple belt reversings. Area pressures of up to 8bar, applied by pressurized water cushions, can be realized in this machine at product residence times of several minutes. The pressure can be increased sectionwise through isolated chambers in order to maintain the inlet conditions of the moist sludge and therewith to prevent it from being pushed back. The high intensity press is employed, for example, after ➔ ***decanter centrifuges*** or ➔ ***double belt presses***.

Hindered Sedimentation

One speaks of hindered sedimentation if the particles in a fluid, i.e. a liquid, are not able to settle entirely on their own, but being influenced by each other. Resulting effects start already at a few volume percent ➔ ***solids content*** in a suspension.

Hindered Settling

➔ *Swarm Sedimentation*

HIP

→ ***High-Intensity-Press***

Hollow Fiber Module

Special arrangement of → ***membrane*** in modules for → ***crossflow*** filtration. The membrane consists of hollow fibers with a diameter of less than 1mm, arranged as a bundle in a pipe. These hollow fibers are flown through by the liquid to be concentrated, leaving the module as a → ***concentrate***. The → ***permeate*** flows radially through the membrane walls of the hollow fibers, and is collected inside the enclosing pipe. Hollow fiber modules are utilized for the → ***microfiltration***, but also specially for → ***ultrafiltration***, because they clog easily

Homogeneous

Through and through the same likeness, without distinct places (→ ***isotropic***).

Hop Sack-Weave

→ **Plain Weave**

Hot Filter Press

Combined mechanical-thermal demoisturing process for industrial → ***steam pressure filtration***. It is a special type of → ***membrane filter press***, equipped with alternating heating and membrane filter plates. The heating plates produce a filtrate steam cushion. This steam then displaces mechanically the main portion of the liquid in the filter cake. Following this the cake can be more efficiently thermally contact dried. A similar, however, not as powerful combination of → ***filter presses*** and thermal demoisturing is realized when after a conventional demoisturing step a small amount of heat is transfered and the press chamber is evacuated (e.g. Rollfit of the BERTRAMS company). In this application the drying process takes several hours compared to one hour in the previously mentioned equipment.

Hydraulic Diameter

The characteristic mean → ***pore diameter*** d_h of a → ***bulk***, that would apply if one permeated a bundle of cylindrical capillaries instead of the investigated solid system at hand. Analogous to fluid mechanics the following can be defined for a → ***bulk***:

$$d_h = 4\frac{\varepsilon}{1-\varepsilon}\frac{1}{S_V} = \frac{2}{3\varphi}\frac{\varepsilon}{1-\varepsilon}d_{32}$$

d_{32} = → ***Sauter diameter*** , S_v = specific surface, ε = porosity, φ = form factor. For sand applies for example: $\varphi \cong 1.4, \varepsilon \cong 0.4$ $\Rightarrow d_h \cong 0.33 d_{32}$

Hydrocyclone

Cylindric-conical, non-moving separation device, into which a → ***suspension*** is pumped under pressure tangentially into the cylindrical part. The liquid develops a potential vortex flow, wherein the particles in the flow are exposed to a centrifugal force. Particles up to a certain → ***cut size*** are spun out to the wall and leave the hydrocyclone at the lower, conical end through the → ***apex nozzle***. The major liquid volume flow including the fine particles is

extracted upwardly through a central ➔ ***vortex finder***. These machines are especially suited for the classifying of suspensions. They can be operated with pressures of up to 4bar and cut sizes of from 5 - 500µm. Hydrocyclones are employed for enrichment, ➔ ***thickening***, ➔ ***de-gritting***, ➔ ***classifying***, and desludging. Due to their simple design they can be manufactured readily in extremely different materials of construction as required by the process.

Hydrogen Bridge

Hydrogen molecules for example, due to their extremely strong polar covalent bond, are definite permanent dipoles that strongly attracting each other. These dipole forces are considerably larger than the ➔ ***Van-der-Waals forces***, because their charge distribution is contrary to induced dipoles permanently asymmetrical. Frequently such dipole forces can evolve when a hydrogen atom is bonded with a strongly electron-attracting (i.e. electron negative) atom like F, O, or N. The subsequently positively polarized hydrogen atom acts because of its extremely small size especially strong attracting upon another, negatively polarized atom. The resulting bond is called a hydrogen bridge.

Hydrolysis

(fr. Gk: *hydor* = water, *lyein* = dissolve). Hydrolysis is the splitting of a chemical bond by the addition of water.

Hydrophilic

(fr. Gk: *hydor* = water, *philos* = friend). Water-attracting wetting behavior of water against a solid, where the ➔ ***wetting angle*** has to be $\delta < 90°$. An example for a hydrophilic system is water and glass.

Hydrophobic

(fr. Gk: *hydr-* + *phobia* fear of water). Water repelling-wetting behavior of water against a solid, where the ➔ ***wetting angle*** has δ more or equal 90°. An example for a hydrophobic system is water and Teflon.

Hydrostatic Cake Formation

Phenomena appearing especially at ➔ ***disc filters*** with large disc diameters. Due to the depth of immersion of the disc into the ➔ ***suspension*** a hydrostatic pressure is created, which in combination with the atmospheric pressure on the ➔ ***filter cell*** builds up a ➔ ***pressure difference***, that starts a cake formation on the ➔ ***filter medium*** before the actual vacuum build-up in the cell begins. As the hydrostatic cake formation increases with progressing depth, the commonly occurring problem of uneven cake formation on the filter cells also increases.

Hygroscopic

(i.e. water attracting). Hygroscopic substances (e.g. sulphuric acid, calcium chloride) attract humidity out of the surrounding air.

Hyperbar Filter

Class of continuously working overpressure cake filters, developed in Karls-

ruhe, Germany. Hyperbar filters are characterized by a ➔ ***drum filter*** or ➔ ***disc filter***, which is mounted complete with its drive in a man-sized pressure vessel. There is a manhole for maintenance and service. Detail consideration has to be given to the safe discharge of the moist solids concerning the operation of the system. Hyperbar filters are preferably employed in the mineral, ore, and coal processing industries, as well as in the food and chemical sector. They operate with pressure differences of up to 8bars and can hold more than $100 m^2$ filter area in a single pressure chamber. Hyperbar filters are built by a number of manufacturers, among others by the BOKELA company with their ➔ ***Hi-Bar-Filtration*** technology.

Hyperbar filter with unlocked pressure vessel (view on a BOKELA ***Hi-Bar*** drum filter)

Idle Time

The unproductive time of a ➔ ***discontinuously*** working separation process. This includes e.g. times for the filling with suspension, for the solids discharge and for the preparation of the apparatus for a new filling. The total batch time then results out of the sum of idle time and separation time.

Impact Ring Centrifuge

Continuously working centrifuge with vertical rotation axis and a drum that opens conical upwards and is divided into annular segments for the ➔ **demoisturing** of coarse-grained plastic materials. The product is fed highly concentrated into the lower central part of the rotating drum. The granulated particles are accelerated there and move from one annular segment to the next in the direction of the solids discharge. Upon impact on the respective next ring the adhering water is separated from them and exits through narrow slots in the drum to the outside.

Inclined Plate Clarifier

➔ ***Lamella Clarifier***

Incompressibility

refers to a ➔ ***bulk's*** behavior of not getting compressed by pressure. Incompressibility is exhibited especially with coarser particles in the region over 100µm. Very small particles or ➔ ***flocks***, however, tend to form ➔ ***bulks*** with a distinct ➔ ***compressibility***. Incompressible ➔ ***bulks*** cannot be demoistured by press filters, and the entrained liquid has to be removed by overcoming the capillary pressure by means of gas or centrifugal pressure.

Inert Gas

is a gas concerning the solid-liquid separation that does not react with the mixture to be separated. Generally, this is a nitrogen or helium atmosphere, which requires an enclosing of the separation machine. An inert gas atmosphere is especially necessary if oxidation processes ranging up to explosions by atmospheric oxygen in the air are to be avoided.

Inlet Cone

Form of feed distributor, frequently used in filtering ➔ ***centrifuges***, for the even charging of the ➔ ***suspension*** into the centrifuge drum. If the inlet cone rotates, then it serves simultaneously for the preacceleration of the suspension to the rotational drum speed.

Inner Liquid

is the liquid bound in fine hair cracks or

inside isolated voids of particles in a ➔ ***bulk***. The volume share of this intra-particle liquid is extremely high in biological cells, which for the largest part contain water isolated from the outside by the cell membrane. The inner liquid is not accessible for mechanical demoisturing without particle destruction.

Integrity

is the qualitatively unimpaired condition of a filter element, ensuring a safe functioning of the element in critical filtration processes, such as are demanded in the pharmaceutical industry.

Interfacial Tension

Reversible, isothermal work that is necessary at constant temperature and mole number to enlarge an interfacial boundary surface A by the amount dA. As the molecular attraction forces at the boundary of two immiscible substances (at least one liquid) do not compensate, but form instead a resulting force pointing to the inside of the homogeneous phase, work is necessary to transport additional molecules into the interface. Surface-active substances, also called ➔ ***tensides***, reduce the interfacial tension. It also decreases by a temperature increase. The interfacial tension is measured as force per unit length and is quoted in N/m or mN/m, resp. Water at 20 °C possesses an interfacial tension of 72mN/m. Often the commonly valid term of interfacial tension is synonymously used with the expression "surface tension". However, the latter applies according to its strict definition only to the surface of a substance against a vacuum. In view of measuring accuracy these agree mostly with surfaces against their own respective steam or a gas.

Interior Drum Filter

➔ ***Drum filter*** with the ➔ ***filter area*** located on the interior. The ➔ ***suspension*** is retained in the drum with a flange ring. Such filters are suited for products with a stronger tendency for ➔ ***sedimentation***. Nowadays, they are rarely employed in Europe.

Intermediate Suspensions

Suspension concentration region within the range of ➔ ***hindered sedimentation*** where instabilities of the settling process can appear in the form of ➔ ***channel formation***.

Interstitial Liquid Demoisturing

While it is not possible with a gas differential pressure field to demoisturing the ➔ ***interstitial liquid*** held at the contact points of particles in the ➔ ***bulk*** due to pressure compensation around the contact points, a certain part of the liquid still can be removed by inertia force demoisturing in ➔ ***centrifuges*** operating at extremely high ➔ ***C-values***. The demoisturing of interstitial liquid is characterized as the so-called 4th region of the ➔ ***Bond-curve***.

Interstitial Liquid

remains due to attracting ➔ ***capillary forces*** at the contact points of particles after a mechanical demoisturing pro-

cess. Depending on the geometric conditions and the ➔ ***wettability*** of the solids, the fraction of interstitial liquid held in the ➔ ***bulk*** can reach 8÷20% of the ➔ ***saturation***.

Inverting Filter Centrifuge

A discontinuously working ➔ ***filter centrifuge***, with a drum insert where the ➔ ***filter cloth*** is attached to at one end like a cuff. After the cake demoisturing process, this drum is hydraulically pushed out of the drum in axial direction. This turns the filter cloth inside out and the ➔ ***filter cake*** that had formed on its inside, is now on the outside and can be cast off the filter cloth. An advantage here is the complete removal of the filter cake without a remaining cake layer, as required with ➔ ***peeler centrifuges***. Also, it cleans the cloth extremely well. As the sieve basket rotates in an enclosed housing, the centrifugation can also be superimposed by a ➔ ***pressure filtration*** or ➔ ***steam pressure filtration*** by pressurizing the drum's internal space. This internal space is sealed against the housing by the front plate of the drum insert. The main field of application of this relatively complicated machine is in fine chemicals and pharmaceuticals.

Isoelectrical Point

In ➔ ***colloidal*** ➔ ***ampholytes*** the mobility caused by ➔ ***electrophoresis*** drops to zero and the zeta potential disappears. The ability of suspended particles for ➔ ***agglomeration*** reaches a maximum.

Isokinetic Sampling

refers to taking a fluid sample with minimal interference by the sampling device on the flow. It has to be designed in such a manner that the flow velocity at its intake corresponds exactly to the one of the fluid surrounding the device.

Isotropic

A ➔ ***bulk material*** is defined as isotropic if the center of gravity of each particle has the same probability to be located at any random location in the ➔ ***bulk*** at any random point in time. An isotropic ➔ ***bulk material*** is entirely evenly mixed and does not display any predominant direction.

Kappa-Factor

A connection between cake thickness h_c, filtrate volume V and filter area A can be derived out of a mass balance of the ➔ ***filter cake*** and the ➔ ***filtrate***, aside from this being dependent on the porosity ε and the densities of the solids and the liquid:

$$\kappa = \frac{\frac{m_s}{m_L}\rho_L}{(1-\varepsilon)\rho_s - \varepsilon\frac{m_s}{m_L}\rho_L} = \frac{h_c A}{V} = \frac{c_V}{1-\varepsilon-c_V}$$

As can be recognized out of the formula, the kappa-factor is to be interpreted also as a concentration measure in dependence of the ➔ ***solids volume concentration*** c_V of the ➔ ***suspension***.

Kelly Filter

Old design of ➔ ***leaf filters*** (1905), employed for ➔ ***fine filtration*** and usually works with overpressure. The Kelly filter possesses vertical, rectangular filter blades, which are installed in sequence in a lying pressure tank. For the ➔ ***filtration*** it will be filled entirely with the ➔ ***suspension*** to be separated. The filtrate discharge is performed for each filter element through the lid. After the ending of the filter process the filter container is pulled away from the plate package, which is to be cleaned. The overpressure in Kelly filters lies usually at several bars. At difficult to filter suspensions one can work with ➔ ***filtering aids***.

Kieselgur

➔ ***Diatomaceous Earth***

Kinematic Viscosity

The kinematic ➔ ***viscosity*** ν is expressed in Stokes (1 St = 1 cm^2/sec). It is related to the ➔ ***dynamic viscosity*** η via the fluid density ρ:

$$\nu = \frac{\eta}{\rho}$$

Knife

Device for removal of ➔ ***filter cake*** after the demoisturing step. The knife can actually have a cutting function, such as the ➔ ***peeler knife*** in ➔ ***peeler centriuges***, or only function as a deflecting plate, as in ➔ ***drum filters*** with ➔ ***scraper discharge***.

Konfiltro

A joint development for a ➔ ***belt filter*** by the BAYER company and the BHS company with a heatable and permeable pressing device that is placed on the cake.

Krämer Filter

Combination of ➔ ***crossflow filter*** and

→ ***press filter*** comprising disc shaped filter membranes that are mounted on a rotating shaft in a horizontal pressure tank. Filter cake formation on the → ***filter area*** is prevented through crossflow, so that the → ***suspension*** is highly thickened. After the rotation is stopped the press membranes demoisture the remaining → ***sludge*** extensively. The demoistured solids are subsequently discharged. The field of application for this mechanically relatively complex machine is mostly in the recovery of high value pharmaceuticals or fine chemicals.

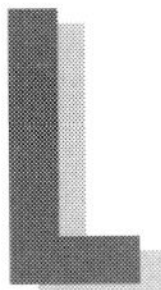

Laboratory Apparatus

Small scale equipment like the → ***Filtratest*** or → ***Centritest*** for simulating solid-liquid separation process in the laboratory. Objective is to obtain data concerning the separation behavior of the → ***suspension*** and the dimensioning of full-scale scale equipment. Since certain machine parameters of the envisioned large scale apparatus cannot be simulated in a laboratory apparatus, like a special agitator machine, testing with a so-called → ***pilot plant*** is frequently conducted on semi-technical scale for proper plant dimensioning.

Lace Weave

→ ***Weave*** in → ***linen*** or → ***twill weave*** where the warp and weft threads, crossing at right angles, have different diameters.

Lambda Value

The term lambda value λ is used in the modeling of the → ***demoisturing*** of → ***filter cakes*** in the centrifugal field. It is defined as a dimensionless kinetic parameter taking into account the cake thickness h_c, the liquid viscosity η_L, the liquid density ρ_L, the → ***centrifugal value*** C, the earth's acceleration g, the mean → ***hydraulic radius*** of the cake pores r_h and the demoisturing time t_2:

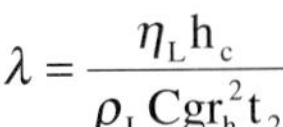

$$\lambda = \frac{\eta_L h_c}{\rho_L C g r_h^2 t_2}$$

Lamella Clarifier

Apparatus for gravitational sedimentation. The clarifying area is increased with tilted, parallel plates separated by only a small distance of several mm up to a few cm between each other. The gap width determines the sedimentation path of the particles on the plate beneath it. The clarified liquid rises over special rising channels and exits on the upper side of the apparatus. If a lamella clarifier is applied as a → ***thickener***, a sludge collecting space beneath the plate assembly is required with sufficient height for the formation of a → ***compression layer***. Lamella clarifiers can be operated in co-current flow (particle and liquid) or in countercurrent flow. In addition, they can be employed in special designs for the separation of a second liquid phase (e.g. oil and water).

Laminar Flow

A laminar flow is defined as the movement of a liquid or a gas, where individual volume elements move past each other without mixing. This flow type is stable only up to a certain → ***Reynolds number***. In the laminar flow of particles in → ***sedimentation*** for example the particle Re number is made up of its settling velocity w, the particle diameter

x, the liquid's density ρ_L and dynamic viscosity η_L as follows:

$$Re_{sed} = \frac{wx\rho_L}{\eta_L}$$

If this value is < 0.2, which is the so-called Stokes' region, the flow is laminar.

Laplace-Equation

The Laplace-equation describes the ➔ ***capillary pressure*** p_c of a system consisting of two non miscible fluid phases (at least one liquid) by linkage of the ➔ ***interfacial tension*** of the liquid γ_L, and both main curvature radiuses of the liquid surfaces R_1 and R_2:

$$p_c = \gamma_L \left(\frac{1}{R_1} + \frac{1}{R_2} \right)$$

For a cylindrical ➔ ***capillary*** with the radius r and the ➔ ***contact angle*** δ results the following:

$$p_c = \frac{2\gamma_L \cos\delta}{r}$$

Lay up

➔ ***Shutdown***

Layer Thickness

Thickness of a ➔ ***bulk*** formed in a separation apparatus. This can be either a ➔ ***filter cake*** or a ➔ ***sediment***.

Leaching Out

➔ ***Extraction***

Leading Edge

➔ ***Forward Edge***

Leaf Filter

Collective name for discontinuously working, cake forming ➔ ***overpressure filters*** with leaf shaped, sequentially arranged filter elements with up to several square meters of filter area. The filter leafs are made out of several layers of wire cloth. The outer layer is a closely meshed, woven ➔ ***filter cloth***, the middle layer a coarse ➔ ***weave***, in order to drain the filtrate. Leaf filters are usually operated with pressures of up to 6 bar and used in the separation of difficult to filter ➔ ***suspensions***. The solids discharge can be carried out either as an extremely thickened suspension or a ➔ ***filter cake*** breaking in lumps. Leaf filters are also being operated as ➔ ***fine filters*** with ➔ ***precoat*** as well as ➔ ***deep bed filters***. A well-known example for a leaf filter is the so-called ➔ ***Niagara filter***.

Leakage Air

➔ ***Secondary Air***

Leaving Filter Belt

A ➔ ***filter cloth*** for the discharge of a filter cake from a ➔ ***drum filter*** that is guided away from the drum around a small diameter spindle. The cloth returns to the drum via tension and deflection rollers. A running cloth is typically used for strongly adhesive and cloth conta-

minating ➔ ***filter cakes***, as the sharp deflection around the spindle breaks off the cake. The cloth can be washed subsequently on both sides with high-pressure nozzles before returning to the drum.

Level Controller

Sensor in vertical and horizontal discontinuous filter centrifuges for controlling of the filling process with ➔ ***suspensions***. Generally, the level controller employs a water ski-like sensor, that is mounted on a pivot and that glides on the surface of the rotating drum while pressed on by a spring. The pivoting caused by the increase in fill height can be measured by an electronic position sensor and utilized for the regulation of the filling valve. Objective is to avoid overfilling of filter drum. A novel development by the KRAUSS MAFFEI company combines the basic level indicator with a thermal sensor capable of registering differences in frictional heats between suspension and solids, respectively, and thus is able to indicate when the demoisturing step of the cake begins.

Light Weight Segment

Special designed ➔ ***filter segment*** by the BOKELA company for the ➔ ***Boozer Filters***. The weight reduction from 28 kg down to 18 kg per disc filter segment makes maintenance work easier especially at large diameter disc filters. Compared to a standard filter segment the light weight segment has even a better rigidity.

Linen Weave

➔ ***Plain Weave***

Liquid Bridge

➔ ***Interstitial Liquid***

Liquid Load

A measure for the liquid amount remaining in the solid bulk after separation. Generally, the load B is defined as the ratio of liquid mass m_L and solid mass m_s:

$$B = \frac{m_L}{m_s} \quad (-)$$

Long Arm Centrifuge

Special laboratory ➔ ***beaker centrifuge***, whose beakers are so far extended from the rotational axis that a single average can be assigned for the layer to be centrifuged concerning the centrifugal acceleration.

Machine Parameter

Design dimension characteristic for the geometry of the separation apparatus that influence the process directly. Examples for machine parameters are the ➔ ***cake formation angle*** at ➔ ***rotary filters***, or the drum length of a ➔ ***decanter centrifuge***.

Mashing

The mixing of a particular dry solid into a liquid to produce a ➔ ***suspension***. For example, the suspension prepared for alcoholic fermentation during the wine or beer making is called mash.

Mass Concentration

The mass concentration states the ➔ ***solids content*** in a ➔ ***suspension***. Commonly, the mass concentration is quoted in ($g/l_{Susp.}$), and can be readily determined gravimetrically, but causes problems when comparing suspensions made from different compounds whose densities do not correspond. For these cases the ➔ ***volume concentration*** is better suited.

Mass Throughput

Mass transported or separated in a separation apparatus per time unit. The mass throughput refers mostly to the solids mass throughput. It is often correlated to a ➔ ***filter area***. The area-specific solids mass throughput is stated in (kg/m^2h).

Material Feeder

The purpose of a material feeder is the even distribution across the separating surface of the ➔ ***suspension*** to be separated. Material feeders can be a diffuser like distribution metal plate, groove like distribution facilities with a paddle discharge, and distribution or floating dam plates, respectively. Applications for such devices can be found e.g. in strongly flocculated suspensions in the ➔ ***straining zone*** of ➔ ***double belt presses***.

Mechanical Demoisturing Boundary

The mechanical demoisturing of porous solids systems is subject to certain limits. In the demoisturing of ➔ ***bulk*** solids, the mechanical demoisturing boundary is reached when the ➔ ***coarse capillary liquid*** is removed due to exceeding the ➔ ***capillary pressure*** by the differential gas pressure. Then the liquid in the ➔ ***bulk*** exists only in form of ➔ ***interstitial liquid***, ➔ ***adhesive liquid***, ➔ ***inner liquid***, and isolated liquid regions. In the centrifugal field one can still remove certain portions of these liquids by the acting mass force. In the demoisturing by mechanical pressing, a limit is reached when the solid particles get

destroyed by the compacting pressure.

Mechanical Demoisturing

In the solid-liquid separation technology one principally differentiates between mechanical and thermal demoisturing. While at the thermal demoisturing a phase transition of the liquid into the gaseous state is always included, the mechanical demoisturing is achieved by displacement of the liquid at constant temperature. The mechanical demoisturing occurs under the influence of either the earth's gravitational or a centrifugal field, a hydrostatic head, a hydraulic or mechanical pressure, or a gas difference pressure.

Membrane

(Latin: membrana = skin). Membranes are used in solid-liquid separation in two different manners. For one, impermeable rubber or plastic membranes are employed in the mechanical press demoisturing of sludges in ➔ ***membrane filter presses***. Alternatively, permeable membranes are used as porous ➔ ***filter medium*** in ➔ ***microfiltration*** and ➔ ***ultrafiltration***. The ➔ ***pore size*** is generally located in the sub-µm region. Filtration membranes are offered in a large variety of materials and have to be carefully adjusted to the product to be filtered, in order not to clog too early and thus become inoperable.

Membrane Filter Cloth

A microporous ➔ ***membrane*** for ➔ ***cake filtration*** integrated in a technical ➔ ***weave*** or a ➔ ***fleece*** (patent of the BOKELA company). Aside from producing particle-free ➔ ***filtrates***, the membrane filter cloth has the special advantage of complete suppression of the gas breakthrough during the demoisturing phase which is unavoidable in conventional cake filtration. The principle of this ➔ ***semi-permeability*** lies in the fact that ➔ ***hydrophilic*** ➔ ***membranes*** have to have such small ➔ ***pores*** that their ➔ ***capillary pressure*** cannot be surpassed by the acting gas pressure difference (e.g. vacuum with 0.8 bar). The filter cake on the other hand has to be capable of being demoistured at this ➔ ***pressure difference***. Membrane filter cloths display pore sizes between 0.2µm and up to about 1µm. Cake formation performance and final residual moisture correspond in general to those attained in conventional filtration with a common ➔ ***filter cloth***.

Membrane Filter Plates

Special filter plate for a ➔ ***membrane filter press***, which is equipped with press membranes. These membranes can be hydraulically stretched to the outside to push out the ➔ ***filter cake*** from the filter chambers.

Membrane Filter Press

Further development of a ➔ ***chamber filter press***. In membrane filter presses the ➔ ***filter cake*** can be squeezed in the filter chamber from one side by a press membrane. Filter- and membrane plates alternate with each other. Advantages of this design are that the feed pressure of the ➔ ***suspension*** can be kept low and for this a high pressure pump is not needed anymore, that a

➔ ***residual volume filtration*** can be achieved without difficulties, and last the filter cake can be uniformly compressed.

Membrane Fouling

➔ ***Fouling***

Mercury Intrusion

➔ ***Mercury-Porosimetrics***

Mercury-Porosimetrics

A technique for measuring the ➔ ***pore size distribution*** in porous systems. First, the evacuated voids of a porous system are filled under pressure with mercury. This procedure is also called mercury intrusion. The filling of the ➔ ***pores*** is performed with continuously stepwise increasing pressure. Following each pressure increase one waits until the equilibrium of the mercury-absorption in the specimen is reached. Then, according to the ➔ ***Laplace-equation*** a ➔ ***pore diameter*** can be assigned to each pressure level. The distribution results from the respective quantity of the intruded mercury.

Mesh

Number of openings per linear inch in filter weaves. So, for example 5 Mesh corresponds to a ➔ ***pore size*** of 4000µm; 50 Mesh correspond to 297µm, and 5000 Mesh correspond to 2.5µm.

Mesh Width

Opening cross section of ➔ ***pores*** in a filter weave. The term mesh width is not clearly defined. Mostly, it is perceived as the diameter of a sphere capable of passing through a medium sized mesh. Mesh widths in a technical ➔ ***weave*** are principally size distributed, aspiring a ➔ ***pore size distribution*** as narrow as possible.

Micelles

Molecular aggregates of for example a ➔ ***tenside*** that form on the surface of a liquid which had been prior saturated upon further addition of tenside. If the surface is capable of adsorbing more tensides again for instance through area enlargement, the micelles disappear. The tenside concentration in a liquid at which a micelle formation begins is called the "critical micelle concentration" (cmc).

Microfiltration

A type of ➔ ***surface filtration*** where porous ➔ ***membranes***, generally with a ➔ ***pore diameter*** of less than 5µm, are employed either in ➔ ***crossflow*** or ➔ ***dead end*** mode of operation. Microfiltration is applied in polishing and concentrating of ➔ ***suspensions***, that contain a large amount of submicron size particles.

Microporous

Pore structure with ➔ ***pore sizes*** of 5µm and smaller.

Moisture Measurement

➔ ***Residual Moisture Measurement***

Moisture

→ ***Residual Moisture***

Monofilament

Weave, which is woven from endlessly spun individual threads. Monofil weaves are extremely adjustable in → ***pore size***, and are often applied in the → ***cake filtration***. A lower pore size limit is around 5µm.

Monolayer Filter

Apparatus of the → ***deep bed filtration*** type with a homogeneous bed structure operated as → ***quick filter*** with a highly porous layer of → ***filter aid***. The layer is supported on a perforated horizontal filter floor onto which the feed flows in the direction of gravity. The filter is regenerated by changing the layer or by a momentary flow reversal and → ***back flushing***. Their main application is the field of water purification.

Monoplate Filter

Discontinuously operating filter apparatus for → ***vacuum*** or → ***pressure filtration*** with a typically horizontal filter surface and flowing in the direction of gravity. They usually produce thick cake layers up to 0.5m or even higher from granular, often crystalline substances. They are popular due to their simple design in the laboratory; the → ***Büchner funnel*** as a vacuum filter and the → ***Nutsche filter*** as a pressure filter examples for this. Filtration characteristics derived with these apparatus are quite useful for predicting the performance of other filtration equipment, even continuously working ones.

Monoplate Pressure Filter

Discontinuously operating → ***cake filter*** with horizontally arranged filter area and flowing in the direction of gravity. A driving overpressure is generated by either a pressurized gas or from the head of the → ***suspension*** pumped into the filter chamber. Pressure suction filters are ubiquitous in industry, have less than 10m^2 filter area, and are usually operated with pressures less than 10bar. The thickness of the filter cakes, formed in the monoplate pressure filters, can exceed 0.5m. Monoplate pressure filters with a few cm^2 filter area are popular for laboratory testing experiments concerning the characterization of the filtration behavior of suspensions, and the sizing of → ***cake filters*** (→ ***Filtratest***).

Mother Liquor

A liquid originally contained in the → ***suspension*** to be separated that remains in the → ***bulk*** following the filter cake formation. This mother liquor, however, can be displaced or diluted in a following process step with a → ***washing liquid***.

Multifilament

→ ***Weave*** woven from endlessly spun threads and twisted into a twine. Multifilament weaves due to the thread structure display a certain deep bed filter characteristic which can lead to → ***blockage***, but they are more stable against thread breakage than a → ***monofilament*** weave.

Multilayer Cartridge Filter

Candle shaped filter element flowing from the outside to the inside and working as a ➔ ***deep bed filter***. A multilayer cartridge works like a ➔***bulk multilayer filter***, but with a filter layer made up of several layers of differently fine porous ➔ ***filter media*** (e.g. ➔ ***fleeces***) instead of ➔***bulk ma-terial***. Again the ➔ ***pore size*** decreases in the direction of the flow. Multilayer cartridge filters are used in the purification of liquids with extremely low solids contents (pulp ➔ ***colloids***).

Multi-Pass Test

This test was developed in the first place for the examination of hydraulic liquid filters similar to the ➔ ***Single-Pass Test***. Here the ➔ ***suspension*** is recirculated through the tested ➔ ***filter medium*** in a large number of passes. A particle measuring device registers the amount and size of the particles before and after the filter during the test period.

MWCO

Description for the filtration efficiency in ➔ ***ultrafiltration***. Molecular Weight Cut Off is quoted in (➔***Dalton***).

Needle Felt

→ ***Filter medium*** for surface filtration made of a compacted → ***tangled fiber felt***. The fibers of the → ***fleece***, bedded horizontally at first, are looped by punching needles vertically through the → ***fleece***. Through this felting process the fiber structure receives a greater stability. Needled felts have a random pore structure which can be varied over a wide range as to their retention properties, but exhibit due to their three-dimensional structure a distinct deep-bed behavior, which can lead to irreversible blockage of the pore structure by intruding particles.

Newtonian Flow

→ ***Newtonian Liquid***

Newtonian Liquid

In Newtonian liquids a flow sets in immediately when a force is applied: they have no → ***yield point***. In laminar flow parallel liquid layers slide along each other which generate a shear stress between the layers. The following linear relationship exists between the shear stress τ, the viscosity η, and the velocity gradient (slope) dv/dy across the layers:

$$\tau = \eta_L \frac{dv}{dy}$$

The → ***viscosity*** of Newtonian liquids is constant at all shear strains in contrast to → ***shear thickening*** or → ***shear thinning*** liquids.

Niagara Filter

Discontinuously working → ***leaf filter*** for overpressure operation. The rectangular filter leafs, usually covered with metal weaves, hang vertically on a common header in a vertical pressure vessel. The filtration pressure normally does not exceed 6bar. Niagara filters are well suited for both dry and wet discharge of → ***filter cakes***. Furthermore, they operate as → ***precoat filters*** with a → ***precoat layer*** consisting of → ***filter aids***. Niagara filters possess a wide range of application in many industries.

Nominal Filter Fineness

→ ***Filter Fineness***

Nonionic

Molecules that certainly are soluble in liquids, but do not dissociated as negatively charged anions and positively charged cations.

Nozzle-Type Separator

Special design of a → ***disc stack separator***, named for the solids discharge system. Permanently open

nozzles are installed at the largest radius of the double conical separation space with an opening width adjusted for the specific separation task, through which the settled and highly thickened ➔ ***sludge*** is continuously discharged. The yeast separator used in food processing is a typical example for a nozzle-type separator. In order to safely establish a ➔ ***compression*** *zone* and to prevent a loss of the ➔ ***suspension*** into the sludge discharge, the ➔ ***solids concentration*** in the ➔ ***feed*** material must have a certain minimum value. If this cannot be done a ➔ ***self cleaning separator*** can be chosen as an alternative.

Number of Revolutions

Number of rotations of a centrifuge drum or a filter element of a ➔ ***rotary filter***, generally expressed in rotations per minute (rpm). While the number of revolutions of ➔ ***centrifuges*** can vary from 100 up to several 10,000rpm, it lies in the range of 0.1 to 10rpm for filters.

Nutsche Filter

Simple, discontinuously working apparatus for the ➔ ***cake filtration***. A nutsche filter consists in its basic form of a vessel, closed at the bottom by a ➔ ***filter surface*** that is filled with a ➔ *suspension* for ➔ ***filtration***. Depending on design, the suspension will then be filtered either under ➔ ***vacuum*** or with overpressure in the direction of gravity. Agitators are frequently integrated into nutsche filters to improve the separation, which then are called agitated ➔ ***monoplate pressure filters***. The stirring allows a multitude of additional operations, which can entail a chemical reaction, crystallization, as well as ➔ ***washing*** and drying of the product. Those nutsches are then called ➔ ***filter reactors***.

Operating Parameters

are understood as all ➔ ***variable parameters*** suitable of being altered during the operation, e.g the number of revolutions for a rotating centrifugal rotor, or the ➔ ***immersion depth*** of a filter drum, or the ➔ ***pressing power*** of a ➔ ***filter press membrane***.

Osmosis

Selective passage of liquid components through ➔ ***membranes*** due to a concentration difference. If a liquid permeates a semi-permeable membrane under the force of an outer pressure and the solutes (e.g. ions) are retained on the wall in this manner then it is called reverse osmosis (saltwater desalination).

Overflow

limits the fluid level; for example a ➔ ***peeler centrifuge*** can be operated with an overflow, meaning the ➔ ***suspension*** drains off over the flange ring during the cake formation and the maximal centrifugal pressure acts. In ➔ ***drum*** or ➔ ***disc filters*** the ➔ ***filter trough*** level can be adjusted with a height-variable overflow pipe.

Overflow Weir

Term often used in connection with ➔ ***clarifiers*** and ➔ ***thickeners*** for the overflow rim where the clear liquid evolves. For constant liquid discharge the weir is often toothed or serrated.

Overpressure

Absolute pressure acting against the surrounding atmospheric pressure.

Overpressure Filter

Enclosed filter apparatus, working with an increased gas pressure in relation to the surrounding atmosphere. Overpressure filters work with ➔ ***pressure differences*** up to around 10bar; hydraulic or mechanical pressure filters are employed for larger pressures.

Oversize Particles

have a considerably larger diameter than the average solids and appear sporadically as a contamination in a ➔ ***suspension***. Oversize particles often have to be separated before the actual separation process, in order not to impair the function of the following separation equipment.

Oyster Filter

The Oyster Filter is a special variant of the BOKELA ➔ ***Hi-Bar Filtration*** technology especially developed for the ➔ ***filtration***, ➔ ***wahshing*** and ➔ ***demoisturing*** of ➔ ***suspensions*** in the

chemical, pharmaceutical and food industry. The Oyster Filter is mounted in a shell-like opening pressure vessel which provides for a good accessibility of the filter. A special feature of this innovative pressure drum filter are the individually exchangeable → ***drum filter*** cells.

The BOKELA ***Oyster Filter*** with steam cabin

Paddle Washer

developed by the BOKELA company for ➔ ***countercurrent washing*** of granular substances.

BOKELA ***paddle washer***

The material to be washed is fed at the low point of a sloped flow channel, which has semicircular troughs in the bottom. Slowly rotating paddles mix the suspension and convey the solids up the channel. The washing liquid is added at the high point of the channel for countercurrent flow. The solids are extensively mixed with the washing liquid and conveyed from cavity to cavity upward the channel, The washing liquid leaves highly enriched the apparatus at the lowest point. The washing can be controlled by adjustments to the channel slope, the number of paddle revolutions, and the solids and wash liquid feed rates, respectively.

Pan Filter

➔ ***Table filter***

Parallel Connection

Parallel arrangement of multiple units of processing equipment when an individual unit is too small to process the total feed, or when quasi-continuous operation has to be realized in the case of discontinuously working units. For the latter, each individual machine is operated in a time-shifted manner for continuous processing of the feed stream.

Particle

Equivalent term for solid particles of small dimension in the mm or µm range.

Particle Fraction

Particle collective with a defined property. This can refer to either the particle type or a certain particle size range.

Particle Size Analysis

Measurement of a ➔ ***particle size distribution***. There is a host of different techniques in use that can vary in their ➔ ***dispersity charcterstics*** as well as in their ➔ ***quantitative aspects***. An example would be a sieve analysis where the dispersity characteristic is the sieve hole diameter and the quantitative aspect is respectively the volume or the mass of a particle fraction.

Particle Size Distribution

Result of a ➔ ***particle size analysis***. The particle size distribution is expressed either as sum or density distribution. The particle sum distribution Q(x) states how many percent of the total particle amount are smaller than the examined particle size x. At the maximum particle size x_{max} the particle sum is assigned the value $Q(x_{max}) = 1$. At the minimum particle size x_{min} the particle sum has the value $Q(x_{min}) = 0$. The particle density q(x) is the particle amount in a differential particle size interval dQ(x) over the differential particle size interval dx. Both distributions are related as follows:

$$q(x) = \frac{dQ(x)}{dx}$$

Particle Size

refers to a geometrical particle dimension. Often a definite description is impossible due to an irregular particle shape and one uses a so-called statistical particle diameter or an ➔ ***equivalent diameter***.

Particle Size Measurement

➔ ***Particle Size Analysis***

Pattern Repeat

Term applied in characterizing a ➔ ***woven cloth***. The pattern repeat entails the minimum number weft and warp threads cross each other in an uninterrupted sequence when arranged side by side in the weft- and warp direction.

Peeler Centrifuge

Discontinuously working ➔ ***filter centrifuge*** with a horizontally rotating axis. Its name is derived from the solids discharge knife that touches the ➔ ***filter cake*** during the drum rotation. This peels the cake off the drum down to a ➔ ***residual product layer***, that is left on for the protection of the ➔ ***filter medium***. The solids are ultimately discharged via a ➔ ***conveyor chute*** or a ➔ ***conveyor screw***. Additionally, peeler centrifuges can perform intensive ➔ ***cake washing***. These centrifuges have diameters of approximately 0.5m to 2.0m, can operate at ➔ ***C -values*** of approx. 400 ÷ 3000, and have a throughput of approx. 0.1 ÷ 50m.t./hr.

Peeler Knife

Cake removal device in ➔ ***peeler centrifuges***. A standard peeler knife reaches across the entire depth of the drum; alternatively, a short knife with axially oscillating movement is used. The peeler knife is either radially pivoted against the cake on the rotating drum or vertically driven against it. To protect

the ➔ ***filter medium*** against damage a thin product layer remains on the drum. The cake discharge with peeler knives at ➔ ***centrifuges*** is roughly equivalent to the ➔ ***scraper discharge*** used in ➔ ***drum*** *and* ➔ ***disc filters*** if reverse pneumatic pulsing is not considered.

Peeling Disc

Stationary double disc with curved flow channels in the gap leading to the center of the disc. The rotating liquid surrounding the disc enters into the flow channels and drains through the center of the disc at a high velocity. Such discharge devices are found in ➔ ***disc stack separators*** and ➔ ***decanter centrifuges*** as well as ➔ ***centrifugal mixers***.

Peeling Pipe

used in ➔ ***centrifuges*** to transfer clarified liquid from the process chamber. The peeling pipe dips into the rotating liquid and as the kinetic energy of the liquid is sufficiently high an external pump is eliminated. Hence a peeling pipe or a ➔ ***peeling disc*** is also dubbed a centrifugal pump.

Peeling Pneumatics

(also known as the Titus-System) is an aspirating pipe used in ➔ ***peeler centrifuges*** for the solids discharge. Located directly behind the ➔ **peeler knife** this pipe pneumatically takes the peeled solids out of the centrifuge drum. The peeling pneumatics can be especially beneficial for subsequent thermal drying operations.

Pendulum Centrifuge

Discontinuously operating ➔ ***filter centrifuge*** with a vertically rotating axis. The drum is rigidly mounted to an oscillating housing, supported by three telescope-type spring legs. Therefore a special foundation for the absorption of vibrations is eliminated. Solids discharge can be performed in conformance to product requirements, e.g. ➔ ***peeler knife***, ➔ ***peeling pneumatics***, detachable filtering bags, or manual removal. Pendulum centrifuges work in the rpm range of 200 to 3,000, and have drum diameters of 0.3 to 2.0m. Suspension throughputs of approx. 0.1÷20m.t./hr can be realized. They are easy to clean, flexible in operation, and therefore well suited for a quick product change.

Perlite

are fused, ground, and classified swelling clays of volcanic origin. Their structure is leaf shaped and can be chemically grouped as silicates of sodium, potassium, calcium, or aluminum. They are extracted from open pit mines. Perlites are used as ➔ ***filtering aids*** to render dense bulk structures more porous.

Permeability

is the flow resistance of a porous system, termed ➔ ***cake permeability***. The specific permeability of a ➔ ***bulk*** is independent from the layer thickness and determined by the geometry of the pore system. It can be measured in a filter test via the ➔ ***Darcy equation***.

Permeate

The clear liquid produced by ➔ ***microfiltration*** and ➔ ***ultrafiltration***; equivalent terms are ➔ ***filtrate*** in filter apparatus and ➔ ***centrate*** in centrifuges.

pH Value

Negative common logarithm of the hydrogen ion concentration used to quantify the hydrogen or hydroxyl ion concentration in aqueous solutions. pH-values of < 7 are described as sour or acidic, pH-values > 7 as alkaline or basic. At the pH of 7 a solution is neutral, because the concentrations of hydrogen and hydroxyl ions are equal.

Pilot Plant

Experimental equipment for testing on a semi-technical scale. A pilot plant should be as small as possible in order to minimize the operational efforts, but large enough to provide realistic data for the design of a full scale machine. Especially important is the investigation of operational parameters, such as suspension homogenization or solids discharge behavior.

Pipe Module

Membrane filter medium assembled in a pipe of several mm or cm diameter for ➔ ***crossflow filtration***. They offer in comparison to other module designs only a small membrane area per module volume, but are well suited for applications in ➔ ***microfiltration*** as they do not make high demands on the pretreatment of feed ➔ ***suspensions*** to be processed due to their relatively large cross-sectional flow area.

Piston-Spring Model

is used conceptually for describing the ➔ ***consolidation*** of ➔ ***compressible*** ➔ ***bulk materials***. Accordingly, if a bulk is compacted by pressing the structural pressure and the liquid pressure change inversely in the bulk. The structural pressure rises from zero to the level of the pressing force, whereas the liquid pressure declines from the value of the pressing force at the beginning of the consolidation process down to zero at the equilibrium state.

Plain Weave

Special pattern for a ➔ ***weave*** representing the tightest and strongest interlacing of warp and weft threads. In earlier times, it was named calico or muslin weave for cotton, cloth weave for wool, and taffeta in silk weaving, respectively. Weaves in plain weave texture show the same appearance on both sides as one weft thread interlaces a warp thread. By interweaving two or more weft threads with two or more warp threads a rib- or basket weave is formed.

Pleating

Accordion-style folding of a ➔ ***filter media*** for ➔ ***cartridge filters***. Through pleating a filter cartridge with a large surface area is produced, enabling in return large throughput performances with a low starting pressure loss.

Plug Flow

in a ➔ ***bulk*** is, in contrast to ➔ ***fingering***, characterized by a uniformly progressing liquid front. One differentiates in the modeling of the displacement of liquid in a ➔ ***capillary*** by centrifugation between plug and ➔ ***film flow***.

Plunger

Accessory for ➔ ***bag filters***. Plungers are inserted in a bag to reduce dead space (e.g. glue filtration). They are hollow bodies, which are adjusted in their size to the bag's interior to reduce the loss of product.

Police Filter

Generally, a discontinuously operating filter located downstream of a separation process for trapping particles that inadvertently can appear in the primary filtrate. ➔ ***Candle*** and ➔ ***bag filters*** are typically used as police filters.

Polishing Filter

➔ ***Police Filter***

Polymeric Flocculant

High molecular-weight polymers, such as polyacrylamides, with molecular weights in the magnitude of 10^7g/mol. One classifies according to their dissociating groups between cationic, anionic and nonionic polymers. Polymeric flocculants cross-link particles in a ➔ ***suspension*** that form ➔ ***flocks*** which enhances their separation in settling as well as in filtrating processes. Polymeric flocculants are offered either as solid granulates or liquid concentrates and have to be conditioned prior to use.

Pore Diameter

➔ ***Pore***

Pore Flow

relates to the flow of a fluid in the ➔ ***pores*** of a ➔ ***filter medium*** or a ➔ ***bulk***. The ➔ ***empty pipe velocity*** of the pore system is calculated as the product of the mean pore flow velocity and the mean ➔ ***porosity***. Known equations for describing the pore flow in ➔ ***bulk materials*** are respectively the ➔ ***Carman& Kozeny equation*** and the ➔ ***Gupte equation***.

Pore

In solid-liquid separation a pore describes the void space surrounded by solids in a ➔ ***filter media*** or in ➔ ***bulk materials***. The geometry of such pores is often quite complex and exact description can not be given. Hence, the pore size, i.e. the pore diameter, is often represented by substitute values such as the circle equivalent to the pore cross sectional area. A measuring method for the determination of pore sizes in filter media is the ➔ ***bubble point test***.

Pore Size

➔ ***Pore***

Pore Size Distribution

The pores in a ➔ ***filter media*** or in

➔ ***bulks*** are always more or less a size distributed that can be approximated in the same form as a ➔***particle size distribution***, as a pore sum distribution, or pore density distribution as a function of the ➔ ***pore size***.

Porosimetrics

Measurement techniques for the ➔ ***pore size*** or the ➔ ***pore size distribution*** in ➔ ***filter media*** or ➔ ***bulks***. In ➔ ***mercury porosimetrics*** the evacuated structure of a porous system is successively filled. With the bubble point procedure a porous body filled with a ➔ ***wetting*** liquid is stepwise demoistured by application of a steadily increasing gas pressure differential. Both procedures utilize the connection between ➔ ***capillary pressure*** and ➔ ***pore size*** given by the ➔***Laplace equation***.

Porosity

The porosity ε of a ➔ ***bulk*** is defined as the ratio of void volume V_v and total volume V_{tot} of a ➔ ***bulk***, which in turn is the sum of void- and solids volume V_s:

$$\varepsilon = \frac{Vv}{V_{tot}} = \frac{Vv}{Vv + Vs}$$

The porosity is directly connected to the ➔ ***void ratio e***:

$$\varepsilon = \frac{e}{1+e}$$

Potential Vortex

Flow form (also known as ➔ ***Helmholtz vortex***) which develops in ➔ ***hydrocyclones*** within the so-called ➔ ***primary vortex***. The rotational velocity increases radially inwards up to a maximum. For such a flow the momentum equation based on the tangential velocity v and the radius r yields the following relation for different radii:

$$v_1 r_1^m = v_2 r_2^m$$

For frictionless flow the exponent m equals 1. For the frictional flow of suspensions in ➔ ***hydrocyclones*** a value of m = 0.5 is often used.

Precipitation

Solidification of dissolved compounds by adding a suitable substance (i.e. precipitation agents). The insoluble solid precipitates can then be removed by a solid-liquid separation.

Precoat

➔ ***Precoat filtration***

Precoat

➔ ***Precoat Layer***

Precoat Filtration

Another term for ➔ ***clarifying filtration*** where a coarse porous layer of a filter aid is filtered first on a cake-forming ➔ ***vacuum*** or ➔ ***overpressure filter***. Through this auxiliary layer the actual suspension is then filtered. Aside from ➔ ***surface filtration*** there is also a ➔ ***deep bed filtration*** component. Generally, ➔ ***precoat filtration*** is applied when a clear liquid product with as few particles as possible is needed.

Important fields of application are in the beverage and food industries, as well as in biotechnology. Discontinuous ➔ ***candle filters***, ➔ ***leaf filters***, as well as continuous ➔ ***drum filters*** are employed. The ➔ ***disc stack separator*** and the ➔ ***crossflow filter*** equipped with membranes are competing separation options.

Precoat Layer

A filter layer, formed by ➔ ***cake filtration*** from a ➔ ***filter aid***, through which the process suspension is filtered during a ➔ ***precoat filtration***. Precoat materials are ➔ ***diatomaceous earth***, ➔ ***activated carbon***, ➔ ***perlite***, wood flour et al; they usually offer a large specific surface for stopping large amounts of pollutants. Aside from the purely mechanical retention, ➔ ***adsorption*** can be utilized to remove compounds that are dissolved or form ➔ ***colloids*** in a liquid

Pre-Filter

Apparatus employed for screening of ➔ ***oversize grain*** to protect the subsequent separation equipment. Pre-filters are installed upstream of ➔ ***disc stack separators*** and ➔ ***hydrocyclones*** to prevent potential blocking of their discharge nozzles.

Press Belt

employed on ➔ ***drum filters*** or ➔ ***belt filters*** to cover partially the demoisturing section of the ➔ ***filter cake*** for mechanical compression. The press belt, usually designed as a rubber belt, is guided over a roller system to the filter cake, where it is pressed on by press rollers. As long as the press belt seals on the edge of the filter cake, the existing gas pressure differential can also be utilized for pressing, thereby acting not only linearly but also uniformly over the entire surface. Such a pre-press system is of advantage when the filter cake has a tendency for ➔ ***shrinkage crack formation*** as it can be stabilized by a certain pre-compression.

Press Filter Automat

Special design of a ➔ ***membrane filter press***, with a vertically plate package and horizontal filter chambers. The ➔ ***filter cloth*** is continuous that zig-zags through all chambers. After the opening of the plate package the cloth is transported by one plate length forward, and the cakes are safely discharged to the sides of the chambers. Filter cakes can be washed, pressed and demoistured with gas pressure. Press filter automated machines are build with up to 150m^2 filter area and work with gas pressures of up to approx. 6bar. The horizontal position of the filter plates promotes the consistent product distribution in the filter chamber as a well as the efficient ➔ ***washing*** of the cake.

Press Filtration

A process, where the ➔ ***filter cake*** after formation is either compressed by the ➔ ***suspension***, or undergoes additional pressing (➔ ***filter press***), or by a mechanically applied pressing power (➔ ***membrane filter press***), for subsequent demoisturing. Depending on the filter design, pressing powers of a few bars up to high pressures of 150bars are

applied. Press filtration is obviously applicable only if the filter cakes display a distinct ➔ ***compressibility***. This is especially the case with extremely fine grained ($x<10 \mu m$), non-rigid (organic substances), or strongly flocculated particles. This ➔ ***compression*** is also called consolidation.

Press Roller

Device to apply pressing power on demoistured ➔ **filter cakes**. In ➔ ***double belt presses***, the cake which is held between two filter belts moves through a narrow slot formed by rollers facing each other, alternatively a pressure is created by winding the belts around a roller. On ➔ ***drum filters*** and ➔ ***belt filters***, press rollers are installed as an additional facility for the post-demoisturing of the cake during the vacuum filtration. Often they are combined with a ➔ ***press belt***, which they press onto the cake.

Pressure

Physical parameter, defined as the quotient of a force and the area, on which this force acts against the area's normal line. The pressure is expressed in the following units: $1 Pa = 1 N/m^2 = 10^{-5}$ bar $= 7.5\ 10^{-3}$ mmHG (Torr) $= 1.45\ 10^{-4}$psi $= 9.869\ 10^{-6}$atm.

Pressure Difference

Notion in solid-liquid separation for the acting pressure difference across a ➔ ***filter medium***, representing the driving potential for the separation of a ➔ ***suspension*** that is supplied by an outside pressure source.

Pressure Drop

➔ ***Pressure Loss***

Pressure Filter

Filters where the driving potential is a pneumatic or hydraulic pressure above atmospheric pressure. Pressure filters are employed as ➔ ***cake*** or ➔ ***deep bed filters*** for both discontinuous and continuous operation. The term is physically not clearly specified, as for example ➔ ***crossflow filters*** are usually not characterized as pressure filters.

Pressure Leaf Filter

➔ ***Leaf Filter***

Pressure Loss

The loss of pressure caused by fluid friction in a permeated system, such as a pipe or a ➔ ***filter cake***.

Pressure Nutsche Filter

➔ ***Monoplate Pressure Filter***

Pressure Rotary Filter

➔ ***Hyperbar Filter***

Pressure Siphon Peeler Centrifuge

➔ ***Siphon Peeler Centrifuge***

Pre-Thickening

The pre-treatment of a suspension, where as much as possible of the particle-free liquid is removed from a

diluted ➔ ***suspension*** **by relatively** simple means, in order to reduce the burden on the often considerably more complex equipment for the following mechanical demoisturing, or to allow their application in the first place. For instance, ➔ ***Pusher centrifuges*** require a certain minimum feed concentration for proper operation, i.e. to prevent an excessive increase of the ➔ ***solids loss*** through the sieve as well as to be able to form and demoisture the ➔ ***cake*** in the relatively short residence time. Gravitational ➔ ***thickeners***, ➔ ***decanters***, ➔ ***thickening filters***, and ➔ ***cross-flow filters*** are used as thickeners. The thickening process is often combined with a ➔ ***flocculation*** to ease the separation of the mostly extremely small particles.

Primary Particles

When particles in a ➔ ***suspension*** are present as solid ➔ ***agglomerates***, then the originally individual particles that make up the agglomerate are called primary particles.

Primary Vortex

Flow in a ➔ ***hydrocyclone*** developing immediately behind the suspension inlet. At the opposite end of the hydrocyclone, at the throttled ➔ ***apex nozzle***, the flow direction is reversed, and the liquid leaves the cyclone in a secondary vortex through the ➔ ***vortex finder***. Coarse particles in primary vortices are separated onto the outside at the cyclone wall, while the fine particles follow the liquid flow on the inside. The primary vortex is a ➔ ***potential vortex***, meaning the flow velocity increases towards the inside. The maximum flow velocity is attained at the radius of the vortex finder, where also the ➔ ***cut size*** is determined.

Product Moisture

Amount of liquid which still remains in the separated solids after the solid-liquid separation. It is reported, respectively, as ➔ ***residual moisture***, as ➔ ***dry substance content***, or as ➔ ***saturation degree***.

Product parameters

are intrinsic product properties that influence the separation process. Product parameters are for instance the dynamic ➔ ***viscosity*** η_L of the liquid or its ➔ ***surface tension*** γ_L, the particle size x and the ➔ ***particle size distribution*** Q(x), and the ➔ ***suspension concentration*** c_V.

Pseudoplastic

➔ ***Shear Thinning***

Pull Action Centrifuge

Discontinuously working ➔ ***filter centrifuge***, by the FERRUM company, with a vertical rotation axis and a drum with an open bottom. After filtration and cake demoisturing, the machine is slowed down and the ➔ ***filter cake*** is detached by stretching the ➔ ***filter cloth***. This discharge is downwards out of the machine without a remaining ➔ ***residual layer***.

Pulp

→ *Suspension*

Pulp Density

→ *Suspension Density*

Pusher Centrifuge

Continuously operating, cantilevered filter centrifuge with solids discharge by means of an axially reciprocating → ***pusher plate***. They are offered in single- or multi-stage design with either cylindrical or conic-cylindrical drums. Pusher centrifuges are also suited for an intensive → ***washing*** of the product due to the cantilevered support and the free accessibility of the processing space. A special design features the → ***double acting pusher centrifuge***. Pusher centrifuges as a rule separate particles with diameters larger than 80µm. They operate with C-values of 200÷2500, have drum diameters of 0.15÷1.5m, and can handle suspension streams of 0.5÷100m.t/hr. To assure good filtrate clarity, these machines require feed concentrations of approx. 10÷40%, which is why they are often combined with a → ***pre-thickener***.

Pusher Plate

Axially reciprocating, circular transfer device in → ***pusher centrifuges***. During each forward stroke the previously formed → ***filter cake*** is pushed a step toward the open front end of the centrifuge drum where the filter cake ultimately breaks off.

Quantity Aspect

Term out of the particle measurement technology, describing a type of quantity of ➔ ***particle fractions*** on which a measuring procedure is based. In sieve analysis, for instance, the extracted particle fractions are weighed, thus the quantity type is the mass or the volume. When using the Coulter Counter principle, the particles of each fraction are counted, thus the quantity is the amount number.

Quick Filter

➔ ***Deep bed filter*** with a filtration velocity of approx. 10m/h. These filters operate under a hydrostatic head, form cake layers of approx. 0.5÷2.5m thickness, and use ➔ ***filter aids*** with approx. 0.5÷4mm grain size. Quick filtration is by far the most important among all filtration techniques for the processing of both potable and industrial water, as well as for the ensuing wastewater treatment.

Quick-Start Centrifuge

Discontinuously working laboratory centrifuge (➔ ***Centritest***) characterized by the short time required (i.e.1÷2sec) to reach the set number of revolutions and the subsequent deceleration. The reason for this operational characteristic is to be able to perform realistic design experiments for continuous ➔ ***filter centrifuges*** that have a product residence time of only a few seconds (e.g. ➔ ***pusher centrifuge***, ➔ ***vibratory screen centrifuge***).

Rake

Plow-like transport implement located at the bottom of ➔ ***circular thickeners***, to convey the settled sludge to a sludge discharge opening in the center. The rake rotates at an extremely low speed to prevent a stirring up of settled ➔ ***sludge***.

Radial Control Head

➔ ***Control Head***

Raffinate

Original liquid or ➔ ***suspension*** from which during an ➔ ***extraction*** an ➔ ***extract*** can be made that is enriched with a certain component, often with the aid of an extraction agent.

Rebecel

A type of ➔ ***filter aid*** distributed by the BELLMER company based on cellulose from renewable resources for improved cake drainability. The material is offered in pellets that are admixed after dissolving them in water to a ➔ ***sludge*** before pressing.

Receiver

Commonly, a receiver is perceived as a cylindrical container downstream of a solid-liquid separation process in which gases are expelled from a liquid. In ➔ ***vacuum*** or ➔ ***overpressure filters*** the liquid-gas (air) mixture evolves from the cake demoisturing zone while in ➔ ***centrifuges*** it is the gas (air) carried along with the centrate.

Remanent Moisture

The liquid portion of a ➔ ***bulk*** that cannot be removed any further by mechanical means. Essentially, it comprises ➔ ***interstitial liquid***, ➔ ***ad-hesive liquid***, and ➔ ***inner liquid***. Vacuum or overpressure filtration can also form hydraulically isolated liquid regions in the ➔ ***bulk***.

Remanent Saturation

➔ ***Remnant moisture*** stated as ➔ ***saturation degree***

Re-Mashing

➔ ***Mashing***

Re-Moisturization

Phenomena in filters observed during the cake removal step, when the driving ➔ ***pressure difference*** is shut off. Some of the liquid remaining in the structure of the ➔ ***filter medium*** or in voids of the ➔ ***filter cell*** can be sucked back by capillary action in the demoistured filter cake.

Residual Moisture

Mass-related definition of the liquid content in a ➔ ***bulk*** following a separation process. The determination of the residual moisture (RM) is simply performed by respective weighing of the moist and dry cakes. The mass of the liquid m_L is then related to the total mass of the moist ➔ ***bulk*** m_{tot}, i.e. the sum of solids mass m_s and liquids mass m_L. The residual moisture is stated as (mass %). The residual moisture of materials with different densities cannot be compared with each other:

$$RM = \frac{m_L}{m_s + m_L}.$$

Residual Moisture Measurement

determination of the ➔ ***residual moisture*** *of* moist materials (e.g. ➔ ***filter cake***, ➔ ***sediment***, ➔ ***bulk***) is either performed off-line or on-line. A representative sample is gravimetrically analyzed, i.e. the moist material is weighed, then dried, and weighed again, in the off-line technique, while in on-line determination the residual moisture is measured directly in the product flow. It should be mentioned, that both the microwave technique (i.e. integral residual moisture) as well as the infrared absorption process (i.e. measurement of the surface moisture) have proven reliabilities.

Residual Product Layer

In a number of filter types, such as the ➔ ***table filter*** or the ➔ ***peeler centrifuge***, the ➔ ***filter cake*** shall not be removed completely, because a mechanical cake discharge device like a ➔ *screw or* a ➔ ***peeler knife*** can destroy the ➔ ***filter medium***. Therefore a several mm thick product layer is left on the filter medium. Over the course of a number of filtration cycles, this can clog the medium with fine particles and render it impermeable, so it has to be regenerated periodically by ➔ ***back flushing*** or removed. The existence of the residual product layer is of advantage for the extensive prevention of ➔ ***turbidity*** at the beginning of the cake forming, since it acts as an additional filter medium.

Residual Volume Filtration

In discontinuously working ➔ ***candle filters*** or ➔ ***leaf filters*** the problem arises of how to process the residual suspension volume left in the lower vessel section after the main filtration phase is completed. One possibility is to let it drain off and filter it with the next batch. Another one is to filter the heel, i.e. the residual volume, by spraying it onto the already formed ➔ ***filter cake*** via a recycle loop.

Resistance Force

The sum of pressure and friction forces exerted by a fluid flow on the surface of a particle. The component in flow direction is called resistance force and the one perpendicular to it as dynamic ➔ ***buoyancy***.

Re-suspending

of a ➔ ***filter cake*** by addition of liquid

and application of stirring energy. Applications for this can be found in → ***dilution washing*** where a yet contaminated filter cake is resuspended in a pure → ***washing liquid*** and filtered anew.

Retentate

→ *Concentrate*

Revamping

Revamping of filter plants is carried out to increase the filter capacity, to reduce the residual moisture of the filter cake or to improve the handling and availability of filters. According to a special optimization program developed by BOKELA company filters like → ***disc,*** → ***drum,*** → ***pan,*** → ***belt filters,*** → ***filter presses,*** → ***Kelly filters,*** → ***Niagara filters,*** etc. can be upgraded in three steps including first test trials to determine the optimization potential up to the start up of the modified filter.

Reverse Osmosis

→ *Osmosis*

Reynold's Number

Most important dimensionless number in viscous flow named after the English physicist Reynolds. It is the ratio of inertial to frictional forces, and contains the flow velocity v, a characteristic length l, and the → ***kinematic viscosity*** ν:

$$\mathrm{Re} = \frac{\mathrm{vl}}{\nu}$$

Rheopexy

is a non - → ***Newtonian flow behavior***. The viscosity of a rheopex solution increases with increasing duration of shear forces acting on it, i.e. the fluid becomes thicker.

Richardson & Zaki Equation

Equation describing the settling velocity u of the suspensions separation zone in → ***hindered sedimentation*** where the → ***particle size*** and shape lose influence on the settling rate and the → ***concentration*** of particles c_V (volume concentration) becomes the decisive factor:

$$\mathrm{u} = \mathrm{u}_{\mathrm{St}}\left(1 - \mathrm{c}_{\mathrm{v}}\right)^{4.65}$$

Here u_{St} is a fictitious settling velocity of a representative, average size particle. The actual suspension settling velocity decreases rapidly with increasing concentration.

Rip Weave

→ *Plain Weave*

Rising Channel

A vertical bore hole through the → ***disc package*** of → ***disc stack separators***, used for separating immiscible liquids with different densities. The positioning of the channels depends on the volume ratio of the components to be separated, and the degree of clarity required of either component.

Roller Discharge

Special type of filter cake removal at

➔ ***drum filters***. It is applied with sticky, dough-like, and pasty ➔ ***filter cakes*** (e.g. red mud in bauxite processing), where a ➔ ***scraper discharge*** with ➔ ***compressed air repulsion*** would not be successful. The filter cake is taken up by a roller of small diameter, rotating opposite to the drum and pressing against the filter cake, which then can be cut away by a knife. Often a toothed comb is utilized instead of a knife, in order to give the cake remaining on the roller a jagged structure. This leads to an especially good connection with the cake to be newly removed.

Rollfit

➔ ***Hot Filter Press***

Rotary Filter

➔ ***Continuously working filter*** with the filter elements rotating at a specified ➔ ***number of revolutions*** on a hollow ➔ ***filter shaft*** through which generally the ➔ ***filtrate pipes*** exit. Typical examples for this are ➔ ***drum filters***, ➔ ***disc filters*** and ➔ ***table filters***. The rotary filters also include the ➔ ***belt filters*** that have a horizontal filter belt leading running around two shafts.

Rotary Siphon Cup

A special ➔ ***rotary cup*** in ➔ ***peeler centrifuges*** with a radius a few cm larger than the ➔ ***filter medium***. This produces a liquid head of several cm height behind the filter medium. Its suction generates a ➔ ***vacuum*** behind the filter medium up to the vapor pressure of the liquid, and a vacuum filtration can superimpose the centrifugation. As soon as the first ➔ ***pores*** of the ➔ ***filter cake*** are de-moistured, pressure compensation occurs and the siphon effect collapses.

Saber Shaped Cell

A special, saber-like design of the ➔ ***filter cell*** of a ➔ ***disc filter***. A frequent problem with conventional ➔ ***disc filters*** is an uneven ➔ ***cake thickness***, resulting from the differences in retention time of the ➔ ***suspension*** in individual area elements. To counter this effect, the saber cell is bent at the cell foot in the rotational direction of the disc, so that the cell does no longer emerge out of the suspension solely with the innermost point of the ➔ ***forward edge***, but instead with the entire cell.

Sand Filter

➔ ***Deep bed filter*** use sand of a defined grain size as a filter layer. Sand filters are preferably employed in water purification to treat large volume flow rates. Generally, sand filters work discontinuously and have to be regenerated after reaching the dirt capacity of the filter layer. However, there are continuous sand filter variations where the sand layer is continuously recirculated and regenerated (Dyna-Sand Filter).

Sand Trap

A simple pre-separation tank for wastewater treatment, in which denser sands are first removed by gravitational sedimentation to reduce the burden on the downstream clarythickener, i.e. to prevent blocking of the ➔ ***rake***. The design of these tanks is either circular or rectangular.

Satin Weave

Special design of a ➔ ***weave*** where the weft threads and warp threads cross at right angles in an exactly determined manner. The resulting surface is smooth and without any texture, either on one side (i.e. single satin weave) or on both sides (i.e. double satin weave). The highs and lows of the wefts and warps appear slightly on the surface and are positioned in a diagonal direction without touching each other.

Saturation

➔ ***saturation degree***

Saturation Degree

Normalized characteristic of the liquid content in a porous ➔ ***bulk***. The saturation degree S is the ratio of liquid volume in the ➔ ***bulk*** V_L over the entire void volume V_V

$$S = \frac{V_L}{V_V}$$

If all ➔ ***pores*** in the ➔ ***bulk*** were filled with liquid, then S = 1. If an entirely dry ➔ ***bulk*** is present, then S = 0. The saturation degree of different bulks

made of different substances can be compared with each other as long as the ➔ ***porosity*** of different ➔ ***bulks*** is similar. This is not possible with the mass related ➔ ***residual moisture***.

Sauter Mean Diameter

presents the mean ➔ ***particle size*** corresponding to the specific surface S_V of the entire particle collective. If one would divide the volume of the investigated substance into spheres of uniform size, so that the sum of their surfaces would be as large as the total surface of the collective, then these spheres would have a Sauter mean diameter d_{32}:

$$d_{32} = \frac{\overline{x^3}}{\overline{x^2}} = \frac{6\varphi}{S_V} \quad (\varphi = \text{form factor})$$

Scale-Up

Projecting the results of a ➔ ***laboratory experiment*** or of a ➔ ***pilot experiment*** to full-scale size equipment.

Scoop Pipe

➔ ***Peeling Pipe***

Scraper Discharge

Special type of cake removal at ➔ ***drum filters*** and ➔ ***disc filters***. The ➔ ***filter cake*** is either cut off in the removal section with the scraper working as a knife, or discharged with compressed air and the scraper serves as a deflector plate. In any case, the scraper must keep a certain clearance to the ➔ ***filter medium*** to prevent damage.

Screen Bowl Decanter Centrifuge

features a cylindrical sieve section added behind the solids cone to dewater a granular solid by filtration; e.g. fine coal after it has been preconcentrated out of a diluted ➔ ***suspension*** by ➔ ***sedimentation***. These special designs of a ➔ ***decanter centrifuge*** are built with drum diameters of approx. 0.2÷1.8m, operate with C-values of approx. 300÷6,000, and their suspension throughput can range from 1÷80m.t./hr. A broad particle size spectrum of about 2÷10,000µm can be separated due to the sedimentation prior to the filtration.

Screening Filtration

➔ ***Sieve filtration***

Screw

Discharge or loading device for transferring pasty solids in or out of the process room of a separation apparatus. Discharge screws are central discharge organs of ➔ ***decanter centrifuges***, where they circulate with ➔ ***differential revolutions*** per minute to the main number of revolutions of the centrifuge drum. They are also used in ➔ ***peeler centrifuges***, if the solids cannot be demoistured sufficiently and thus remain sticky and a ➔ ***discharge chute*** would get clogged. Screws have a basic body on which the screw blades are welded on as spirals. The screw blades display a gap or respectively a

screw pitch. They can be equipped with a single or double spiral, left or right handed. The screw pitch can be constant or varies over its length. Screws with a diminishing screw pitch are employed at the ➔ ***worm extruder***.

Screw Blade

➔ ***Discharge Screw***

Secondary Air

can influence the pressure level, leaking unintentionally through a ➔ ***by-pass***, or deliberately through a valve. The quality of the vacuum in filters can be adjusted by the latter in ➔ ***vacuum filtration***. Secondary air flow can increase at the shrinkage crack forming point in filter cakes, so that the vacuum collapses and the filtration process discontinues.

Secondary Vortex

Flow towards the dip tube (➔ ***Vortex Finder***) in hydrocyclones that evolves behind the ➔ ***primary vortex's*** flow reversal at the throttled ➔ ***apex nozzle***. Fine particles in the secondary vortex, i.e. smaller than the ➔ ***cut size***, are carried out with the cyclone overflow.

Security Filter

➔ ***Police Filter***

Sedicanter

Special design of the ➔ ***co-current decanter*** by the Flottweg company for the thickening of dilute and difficult to separate ➔ ***sludges***. Solids discharge is through a double conical drum with a short, steep cone. The settled sludge has to pass a barrier on the way to the discharge driven by hydraulic pressure. This pressure is regulated by the liquid level in the machine and is in turn is adjusted with a ➔ ***peeling disc***.

Sediment

A completely saturated ➔ ***bulk*** of particles formed during a settling process in a gravitational or centrifugal field. The ➔ ***bulk*** must display a greater specific density against the surrounding liquid.

Sedimentation

Settling process of particles in a ➔ ***suspension*** in a gravitational or centrifugal field, if the solids have a higher specific density than the surrounding liquid.

Sedimentation Centrifuge

➔ ***Settling Centrifuge***

Sedimentation Front

Clearly defined region below the ➔ ***clear liquid zone***, in which the entire particle collective settles at the same velocity. It develops readily in ➔ ***hindered sedimentation***, which can occur once a critical ➔ ***suspension concentration*** is surpassed.

Sedimentation Tank

➔ ***Gravitational Thickener***

Sedimentation Velocity

Rate at which an individual particle or

a ➔ ***sedimentation front*** settles in fluid. In the laminar flow region the sedimentation velocity of individual particles can be calculated via ➔ ***Stokes' law***, whereas the ➔ ***Richardson & Zaki-equation*** applies for a ➔ ***sedimentation front***.

Sedimentator

Novel wash process of the BOKELA company and the TICONA company.

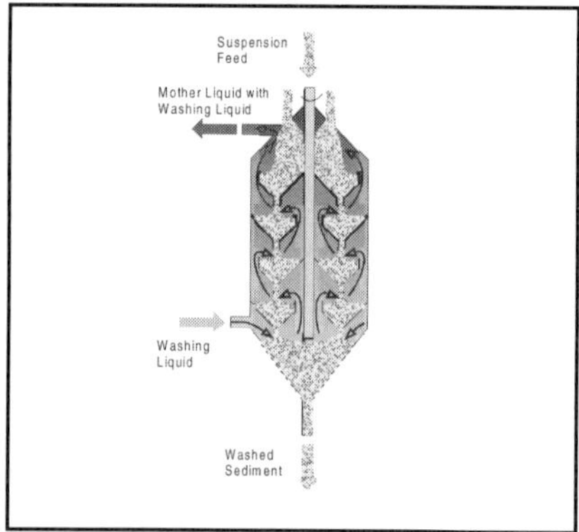

Schematic view of the ***sedimentator***

In the Sedimentator a continuous ➔ ***countercurrent wash*** of solids is realized which is based on an advanced principle of ➔ ***displacement wash***. In the contact zone of solids and washing liquid the settling solids form an expanded bed which enables that the ➔ ***mother liquor*** between the solids is displaced by the ➔ ***washing liquid*** in a highly efficient way leading to high ➔ ***wash degrees*** with a low wash liquid consumption. The novel apparatus is built up like a sedimentation column in a vertical pressure vessel with numerous stages and allows the washing of solids under extreme process conditions (temperature and pressure).

Segregation

De-mixing phenomenon in a ➔ ***suspension***. Often particles get classified according to their ➔ ***particle size***, like for instance when a filtration process is superimposed by an undisturbed ➔ ***sedimentation*** caused by gravitation or centrifugation. The potential for segregation is therefore especially high in horizontal filters and filtrating ➔ ***centrifuges***. Due to this segregation a ➔ ***bulk*** forms with an increasingly finer structured layer. The ➔ ***entry capillary pressure*** in the so-called top ➔ ***clogging layer*** is considerably higher than in a homogeneously structured cake.

Self-Cleaning Separator

Special design of a ➔ ***disc stack separator*** with periodically opened sludge discharge nozzles at the periphery of a dual-conical centrifuge drum. After a critical solids pressure is produced by the settled solids a hydraulic piston slide moves through the settled solids and the discharge nozzles open for a fraction of a second. This discharge type is especially of advantage at low solids concentration (e.g. milk skimming, beverage purification), whereas the ➔ ***nozzle-type separator*** is used for higher solids concentrations.

Self-Suctioning Drum Filter

The pressure difference in ➔ ***drum filters*** is produced by fact that the ➔ ***filter cells*** are connected with the ➔ ***filtrate pipes***.

They lead downwards along the inner wall of the drum and in the circumferential direction against the rotational direction for approx. 1÷1.5m with an open end in the drum's inner space. The filtered liquid causes in such a pipe a hydrostatic head of approx. 0.1÷0.15bar. Self-suctioning drum filters are employed for readily filtering → ***suspensions*** containing fibrous particles, for which a low pressure difference is sufficient.

Self-Transporting Centrifuges

Continuously working, cantilever supported → ***filter centrifuges*** with a conically expanding sieve drum. The driving force for the solids transport is the surface-parallel component of the centrifugal force reduced by the adhesive friction force of the sieve. The control of such machines (e.g. → ***sliding discharge centrifuge***) is difficult, due to the fact that the → ***friction*** depends partially upon the → ***residual moisture*** of the product, thus changing along the path of the cone. These problems can be alleviated through adequate flow guidance (→ ***directed flow screening centrifuge***) or through pulsating acceleration (→ ***tumbler centrifuge***).

Semipermeability

Selective or partial permeability in a separating layer which only allows certain components of a mixture to permeate. Membranes for the → ***ultra-filtration*** pass only molecules of a certain size, which correspond to their → ***MWCO***.

Separation Control Head

→ ***Control valve*** in a → ***cell-less drum filter***. The separation control head forms the transition from the rotating to the stationary part of this filter type. It is employed for the separation of filtrate and suctioned gas (air).

Separation Selectivity

quantifies the loss of coarse grain in the → ***fines*** and of fine grain in → ***coarse material***, respectively. The separation sharpness, i.e the steepness of the → ***fractional grade efficiency curve*** can be defined with characteristic values from the fractional grade efficiency curve as follows:

$$\kappa = \frac{x_{25,t}}{x_{75,t}}$$

$x_{25,t}$ respectively $x_{75,t}$ represent herein the → ***particle sizes*** that are separated at 25% and 75% respectively in the coarse material. In analytical separations the separation sharpness is $0.8<\kappa<0.9$, a sharp technical separation shows $0.6<\kappa<0.8$ and in common technical separations it is lower at $0.3<\kappa<0.6$.

Separator

→ ***Disc Stack Separator***

Septum

→ ***Filter Medium***

Series Connection

→ ***Combination*** of separation equipment arranged in line in order to amplify

or optimize certain separation effects. It can involve similar or different separation machines. A typical series connection is the combination of a ➔ ***thickener*** followed by a ➔ ***filter***. A series connection can also be found at the ➔ ***dilution washing*** with ➔ ***drum filters***, or for the increasing of the ➔ ***cut point*** in ➔ ***hydrocyclones***.

Serrated Weir

Liquid overflow edge with a jagged profile. In this manner an even liquid draining can be achieved over the entire length of the weir edge, even if it is not perfectly horizontal.

Service Life

Time period during which an apparatus or an apparatus element is available for normal operation. It can cause process disturbances and influences the economic efficiency.

Setting Parameters

➔ ***Operation Parameters***

Settling Tank

Circular or square container in which separation by sedimentation of solid particles and liquid occurs under the influence of gravity. The specific density of the dispersed solids necessarily has to be greater than that of the continuous liquid phase. The settled solids are removed from the tank bottom in the form of a thickened ➔ ***sludge***. Ideally, the particle-free liquid is removed at the top of the tank by means of an overflow.

Settling Centrifuge

➔ ***Centrifuge*** based on the separation of the particle solids at an impermeable wall by ➔ ***sedimentation***. Primarily to be mentioned here are the ➔ ***decanter***, ➔ ***disc stack separator***, and ➔ ***tubular centrifuge***. The multiple of the earth's acceleration, ➔ the ***C-value***, attained in these centrifuges ranges usually from a few 1,000 up to several 10,000 in extreme cases, due to the low sedimentation velocity of the extremely small particles. Settling centrifuges are working discontinuously as well as continuously.

Settling Velocity

Rate of settling of solid particles in a liquid under the influence of a gravitational or a centrifugal force. The description of the settling velocity of single particles in the laminar flow region is based on ➔ ***Stokes' law*** while in the sedimentation in concentrated ➔ ***suspensions*** (➔ ***swarm sedimentation***), the settling velocity of the mutually hindering particles can be described with the ➔ ***Richardson & Zaki equation***.

Shear Thickening

Non ➔ ***Newtonian flow***. The viscosity increases with increasing shear stress.

Shear Thinning

Non ➔ ***Newtonian flow behavior***. The ➔ ***viscosity*** decreases with an increased shear stress.

Sheet Filter

A ➔ ***deep bed filter*** employed for the purification of liquids, e.g. beverages, with the outer appearance of a ➔ ***filter press***. Instead of forming a cake in the filter chambers, the diluted ➔ ***suspension*** permeates under pressure the filter sheets, on which contaminants separate. These filter sheets have to be replaced once their absorption capability is exhausted as generally they can not be regenerated.

Shriver-Thickener

Crossflow-filter press that look like a ➔ ***filter press***; the plates, however, are divided in so-called concentrate and permeate plates with a porous separation membrane between. The flow channel spirals from the outer edge of the plates into the center. Along this path an initially diluted ➔ ***suspension*** is thickened.

Shut Down

➔ ***Shut down process*** of a separation machine from an operational state to stand still.

Shut Down Process

Operational phase of a separation machine. One differentiates between a normal turning off and an ➔ ***emergency off***, where the equipment has to be turned off immediately due to a disturbance.

Sieve Filter

➔ ***Sieve Filtration***

Sieve Filtration

➔ ***Surface filtration*** for the purification of liquids contaminated with particle which does not necessarily have to produce a real ➔ ***filter cake***; rather flow is often interrupted at an early stage for a ➔ ***back-flushing***, after attaining a preset pressure loss. A special variant of the sieve filtration is the ➔ ***dynamic sieve filtration*** with the ➔ ***DYNO-Filter***.

Single Filter

Candle shaped ➔ ***sieve filter*** for the cleaning of liquids with low particle contamination. They are typical by-pass filters, because the changing of the filter elements is not possible without flow interruption. A possible solution to this shortcoming are ➔ ***double filters*** or ➔ ***automatic filters***.

Single Particle Sedimentation

Sedimentation behavior of particles in a ➔ ***suspension*** with very low ➔ ***solids concentration***. The particles settle independently without influencing each other. According to ➔ ***Stokes' law***, their ➔ ***settling velocity*** depends on the density difference between solids and liquids, the liquid viscosity, a characteristic particle diameter, and gravitational or centrifugal acceleration, respectively. Single particle sedimentation is especially desired in ➔ ***particle size analysis***, for which actual production samples often have to be diluted.

Single-Pass Test

Determination of the separation ability

of a ➔ ***filter medium*** under realistic conditions where a filter sample passes once. The test filter is fed with a nearly constant particle concentration until a specified maximum ➔ ***pressure difference*** is reached. A particle-measuring device monitors both particle number and size in front and behind the filter.

Siphon-Peeler Centrifuge

Special design of a ➔ ***peeler centrifuge*** in which the ➔ ***centrate*** is not freely ejected into the centrifuge housing, but instead is collected in a ➔ ***rotary siphon cup*** and utilized to produce a ➔ ***vacuum*** behind the ➔ ***filter medium***. The liquid ring in the rotary cup seals the space behind the filter medium against the interior housing, so that an additional pressure filtration can be realized. For this, the pressure gas that permeates through the cake has to be removed by a duct in the centrifuge shaft. A peeler centrifuge of this design is called a pressure-siphon-peeler centrifuge.

Sliding Discharge Centrifuge

Special type of continuously working screen centrifuges. Sliding discharge centrifuges belong to the so-called ➔ ***self-transporting centrifuges***. In these the solids are moved to the discharge solely by the surface parallel component of the centrifugal force surpassing the sum of adhesive and sliding friction force along the conically widening screen basket. Sliding discharge centrifuges have a problematic operation characteristic as the sliding behavior of solids depends among other factors on the properties of solids and liquid, the moisture degree of the ➔ ***bulk***, and the roughness of the screen wall. Sliding discharge centrifuges are generally employed in the separation of highly viscous materials, such as molasses in the sugar industry. Other designs try to control the transport process by built-in devices such as at the ➔ ***directed flow screening centrifuge***, or by periodically changing the centrifugal forces, such as in the ➔ ***tumbler centrifuge***. The sliding process can be influenced especially well if the surface parallel force alone is not big enough for the transport and the friction is surpassed by an axial oscillation, adjustable in amplitude and frequency, of the screen basket like in the ➔ ***vibratory centrifuge***.

Sliding Friction

➔ ***Friction***

Slow Bulk Layer Filter

Sand filter from the field of water treatment. Slow bulk layer filters are based on the principle of ➔ ***deep bed filtration*** and utilize filter layers with more than 1 m thickness. The filter velocity is about 0.05 up to 0.1 m/hr and filter areas reach up to 10,000 m^2. Often a biological reduction process for the organic substance to be separated is combined with the mechanical separation because of the low filtration process.

Sludgy

Highly concentrated ➔ ***suspension*** whose particles are so close that mechanical forces can be transferred between each other, but which are on

the other hand still free-flowing.

Snap-Blow Valve

Auxiliary device for improving the filter cake discharge in ➔ ***disc filters*** and ➔ ***drum filters***. It is essential for a good cake discharge that pressure builds up in the ➔ ***filter cell*** behind the ➔ ***filter medium*** as fast as possible. The inlet cross section at the ➔ ***control head*** opens only very slowly for compressed air especially at low numbers of revolutions. Therefore, a quick-acting valve is located in the compressed air feed pipe that is only opened if the full filtrate pipe cross section is exposed.

Solid-Bowl Centrifuge

➔ ***Decanter***

Solids Concentration

➔ ***Solids Content***

Solids Content

Measure for the amount of solids in a solids-liquid mixture. For a mass-related representation the solids mass m_s is referred to the total mass of the solid-liquid-system $m_s + m_L$ and quoted in (mass %). Depending on whether the solids or the liquid present the continuous phase, this quantity is also described as dry substance content DS or solids mass concentration c_m:

$$DS = c_m = \frac{m_s}{m_s + m_L} 100[\%]$$

Division by the respective densities leads to the term of solids volume concentration c_v:

$$c_v = \frac{V_s}{V_s - V_L} 100[\%]$$

A further relation for the description of the solids content in suspension presents the ➔ ***suspension density***.

Solids Discharge

Every solid-liquid separation apparatus has discharges for respectively the cleared liquid phase and the more or less moist solid phase. In the case of sludge-like solids products, the solids discharge could be a valve or a pump, a ➔ ***discharge screw*** for very moist solids, and in the case of powdery material a ➔ ***conveyor chute***.

Solids Flow

Term for the description of the settling procedures in ➔ ***gravity thickeners*** when ➔ ***swarm sedimentation*** occurs. The solids flow S is defined as the product out of the settling velocity w and solids volume concentration c_v and possesses the dimension (m^3/m^2h). The total solids flow S_{tot} in a gravity thickener consists of the solids flow due to the swarm sedimentation S_{sed} (settling velocity w_{sed}, concentration c_v) and the solids flow due to sludge removal (outlet) from the underflow in the thickener S_u:

$$S_{tot} = S_{sed} + S_u = w_{sed} c_v + \frac{\dot{V}_u}{A} c_{v,u}$$

The sludge outlet velocity V_u is defined as the quotient of the sludge volume stream in the underflow V_u and the thickener cross sectional area A. As concentration and settling velocity both change in opposition directions over the height of a thickener, a critical minimum of the solids flow develops at a certain point. This critical solids flow is the basis for the thickener design. It has to be determined by experiment.

Solids Loss

Amount of solids, passing through the pores of the filter cloth at the very first moment of ➔ ***cake formation*** before a ➔ ***bridge layer*** is formed on it which safely retain the following solids. Material sizes, such as pore width of the filter cloth in relation to the size of the particles to be separated, but also the ➔ ***adjustment parameters***, such as the ➔ ***suspension concentration*** or the ➔ ***pressure difference*** significantly influence the amount of the solids lost.

Solids Mass Concentration

➔ ***Solids Content***

Solids Throughput

Amount of particle solids in a ➔ ***suspension*** that can be processed per time unit by a separation apparatus. The throughput is additionally referred to the ➔ ***filter area*** as the specific solids throughput. For example, the specific throughput of a ➔ ***rotary filter*** is quoted in (kg/m^{2}h).

Solids Volume Concentration

➔ ***Solids Content***

Solution

Homogenous mixture of different substances in which the mutual diffusion and division reaches down to the level of molecules, atoms or ions (true solutions).

Sorting

Segregation of a particle collective into ➔ ***fractions*** of different ➔ ***particle types***.

Spacer

Coarse-sized, meshed insert between two microporous ➔ ***membranes*** employed in the ➔ ***ultra-filtration*** as distance spacers to provide for the unhindered flow of respectively ➔ ***concentrate*** *and* ➔ ***permeate***. Spacers are found for instance in ➔ ***coil*** and ➔ ***spiral modules***.

Specific Surface

Surface area of a particle typically relating to either volume or mass of the solids. The volume related specific surface S_v is connected with the ➔ ***Sauter mean diameter*** of a particle collective d_{32} as follows:

$$S_V = \frac{6}{d_{32}}$$

Spiral Module

Filter element design used in the ➔ ***crossflow filtration***, especially in the ➔ ***ultrafiltration***. Spiral modules are rolled up alternating layers of ➔ ***microporous membranes*** and ➔ ***spacers***. This creates an extremely large membrane

area per module volume. The feed liquid flows coaxially through the module starting from the front end of the roll. The ➔ ***permeate***, after evolving from the membranes, flows on a spiral-shaped path to the center and is extracted through a core pipe.

Stable Suspension

➔ ***Suspension*** whose particles are prevented from agglomerating by electrostatic repelling forces resulting from surface charges.

Staple Fiber

Fibers (e.g. natural fibers such as cotton) made out of short fiber pieces by twisting. In contrast to ➔ ***monofilaments****, which* are endless and smooth threads, staple fibers possess a greater roughness due to protruding fiber ends. This can increase the separation efficiency of a ➔ ***filter cloth*** woven out of it, however, it possibly complicates the cake discharge due to larger adhesion forces between cake and cloth.

Star Feeder

Device for the discharge of ➔ ***bulk material*** from a pressure vessel. A rotor, divided into individual pockets, rotates sealed inside a horizontal cylinder with openings on the upper- and underside. The pockets are filled from above with product and emptied downwards into the atmosphere. The principle of the star feeder can only be employed for mildly abrasive materials, because otherwise the sealing will deteriorate due to the abrasion.

Starting Process

The start-up of the operation of separation machines from a standstill to stable and stationary operational conditions at the ➔ ***operation point***. Very often the start-up is understood to be the initial taking into operation of a new plant.

Static Buoyancy

➔ ***Buoyancy***

Steam Filtration

➔ ***Steaming***
➔ ***Steam Pressure Filtration***

Steam Cabin

Special designed ➔ ***steam hood*** by the BOKELA company for performing the ➔ ***steam pressure filtration*** process on ➔ ***Hi-Bar Filters***.

Steam Hood

Supplementary attachment on continuous ➔ ***vacuum*** - and ➔ ***pressure filters*** for improving the ➔ ***cake demoisturing*** through ➔ ***steaming***.

Steam Pressure Filtration

Innovative and patented process by the BOKELA company in form of the ➔ ***Hi-Bar-Filters*** for ➔ ***cake demoisturing*** and ➔ ***cake washing***. An externally supplied, saturated or overheated steam is used as the gaseous displacement medium in place of the traditional air. The steam condenses on the cold filter cake surface and forms a sharply

defined and evenly developed ➔ ***condensate front*** which moves through the cake. Thus, the cake is heated up to the condensate temperature and a ➔ ***fingering*** with a premature gas breakthrough is prevented which can occur when air is used as gaseous displacement medium. The special advantage of this process rests on the fact that the largest portion of the filtrate is removed from the cake mechanically, and followed immedi-ately by this a thermal convection drying is performed leading to extremely low moisture contents. Next to demoisturing a ➔ ***washing*** of the cake is also performed, due to condensation and thus a pure liquid input into the ➔ ***bulk***. Up to now the process has been realized on ➔ ***hyperbar filters*** with ➔ ***disc filters*** and ➔ ***drum filters***.

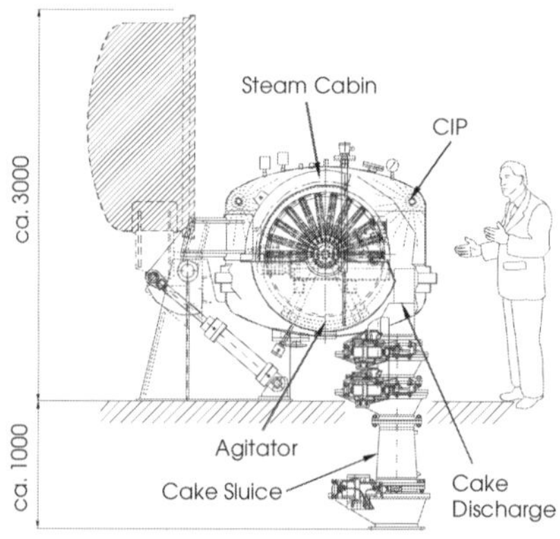

BOKELA ***steam pressure drum filter*** (Hi-Bar Oyster Filter plant)

Steaming

Measure supporting the ➔ ***cake demoisturing*** in the ➔ ***vacuum filtration***, mainly employed for continuous filters, such as ➔ ***belt filters***, ➔ ***drum filters***, or ➔ ***table filters***. Saturated or over-heated steam is applied via a ➔ ***steam hood*** on the previously de-moistured filter cake surface. While the steam is sucked through the cake, it transfers its heat into the cake and condenses there. The heating leads to a reduction of the filtrate viscosity and consequently to an acceleration of the demoisturing process. Decreased ➔ ***product moistures*** are reached within the limited demoisturing time available in continuous filters. In addition to this, the ➔ ***surface tension*** of the ➔ ***filtrate*** is decreased and the ➔ ***capillary pressure*** is slightly lowered. After discharge the heated cake can additionally lose further liquid through post-evaporation.

Sterile Filtration

Separation of all viable microorganisms in the size range of 0.1÷1µm by ➔ ***filtration***. A separation below 0.1µm, e.g. viruses, dissolved toxins, or pyrogenes, is performed by adequate filter beds in an adsorptive manner. The suitability of a ➔ ***filter medium*** as a sterile filter is examined by the ➔ ***bacteria retaining test***.

Stern Layer

Mono- to bimolecular layer of adsorbed ➔ ***counter ions***, solidly bonded on a charged particle surface.

Stirred Pressure Nutsche Filter

➔ ***pressure nutsche filter*** equipped with a stirring apparatus. The stirring allows the execution of a number of ope-

rations in the filtration space, which can range from homogenization of the ➔ ***suspension*** to re-suspending up to solids discharge. The stirred pressure nutsche filter in its most variable form develops into a ➔ ***filter reactor***, in which further operations can be performed, such as chemical reaction, crystallization etc.

Stockpile Demoisturing

Special form of ➔ ***gravity filtration***. The demoisturing of liquid-saturated, granular ➔ ***bulk material*** in a gravitational field is driven by the hydrostatic head of the liquid itself. This pressure has to be larger than the ➔ ***capillary pressure*** acting in the ➔ ***bulk***. During demoisturing the liquid column gets smaller and the hydrostatic pressure decreases, respectively, until at equilibrium of the forces the ➔ ***capillary rise*** of a ➔ ***bulk*** is reached.

Stokes' Law

Equation for the description of the settling velocity v of a spherical individual particle with a diameter x and a density ρ_s in a liquid with a density ρ_L and the ➔ ***dynamic viscosity*** η_L under the earth's gravitational acceleration g, if a laminar flow is present:

$$v = \frac{(\rho_s - \rho_L) g x^2}{18 \eta_L}$$

Particles must be able to settle unhindered by other particles present. If Stokes' law is to be applied onto centrifugation the gravitational acceleration is to be multiplied with the ➔ ***C-value***. It must be verified that ➔ ***laminar flow*** is still present.

Strainer

A continuously rotating, drum-shaped sieve, which is loaded on the inside with for example a strongly flocculated ➔ ***suspension*** for pre-thickening under the influence of gravity.

Straining Zone

Horizontal, pre-demoisturing zone of ➔ ***double belt presses*** in which a crossly flocculated ➔ ***suspension*** is thickened by gravitational demoisturing, so that it can be drawn subsequently into the ➔ ***wedge zone*** between the two filter belts.

Strindlund Filter

➔ ***Self-Suctioning Drum Filter***

String Removal

➔ ***Chain Removal***

String-wound cartridge

Filter element for discontinuous ➔ ***deep-bed filtration*** *made* from twisted yarn (➔ ***stack fibers***), that is coiled to a thick layer around a perforated core from where the filtrate flows. The coiled yarn forms the actual deep-bed filter in whose pores the contaminant particles deposit themselves.

Substitute Cake Thickness

➔ ***Equivalent Cake Thickness***

Suction Filter

General term for the entire class of filters that use a gas difference pressure as the driving potential, generated by the application of a ➔ ***vacuum*** behind the ➔ ***filter medium***. Suction filters are limited in respect to the maximal ➔ ***pressure difference*** by the ➔ ***vapor pressure*** of the liquid.

Support Grain

➔ ***Body-Feed Filtration***

Surface Active Substances

➔ ***Tensides***

Surface Filtration

Filtration where the particles to be separated are retained preferably on the surface of the ➔ ***filter medium*** in contrast to the ➔ ***deep bed filtration***. Surface filtration can be realized as ➔ ***cake filtration***, ➔ ***sieve filtration***, and ➔ ***crossflow filtration***.

Surface Potential

Electric charge of suspended solids particles. Particles in suspensions are often negatively charged. The surface potential is partially compensated by counter ions contained in the liquid, and declines exponentially with an increase in distance from the particle's surface. The surface potential can cause an electrostatic repulsion of particles. If one wants to agglomerate them by utilizing the ➔ ***Van-der-Waals forces*** the surface potential has to be shielded in order to bring the particles to a sufficient proximity.

Surface Tension

➔ ***Interfacial Tension***

Surfactants

Surface-active chemicals tend to gather on ➔ ***interfaces*** and thus do lower the interfacial tension. They mainly consist of a ➔ ***hydrophobic*** group (e.g. hydrocarbon chain with 10÷18 C atoms), or an acrylate group, and a ➔ ***hydrophilic*** group (e.g. -COOMe, -OSO_3Me, SO_3Me, NH_2=NH). According to the polarity of the ionic group one can differentiate between anionic, cationic and non-ionic tensides. Problems are often posed in industrial applications by their limited biological decomposition as well as by a tendency to foam.

Suspension

Mixture of a liquid and a particle-shaped solid. The liquid is in this case the ➔ ***continuous phase**, while* the solid forms a ➔ ***disperse*** or ***discontinuous phase***. At higher solids concentrations the suspension transforms into ➔ ***sludge***, if the particles approach each other so closely that they are capable of exerting mechanical forces on each other.

Suspension Concentration

➔ ***Solids Content***

Suspension Density

The suspension density ρ_{sL} (s= solid, L= liquid) is defined as ratio of mass m

and volume V:

$$\rho_{sL} = \frac{m_s + m_L}{V_s + V_L} = \frac{\rho_s}{c_m + (1 - c_m)\dfrac{\rho_s}{\rho_L}}$$

$$\rho_{sL} = c_v \rho_s + (1 - c_v)\rho_L$$

c_m =➔ ***solids-mass concentration***, and
c_V =➔ ***solids-volume concentration***

Swarm sedimentation

Settling behavior of a particle collective characterized by the fact that size, density and shape of the particles all lose their influence on the sedimentation velocity. A sharp sedimentation front appears with a clear liquid zone above. The main influencing parameter becomes the ➔ ***suspension concentration***. Over the broad range of concentrations where swarm settling is observed some classification effects - especially in centrifugal fields - can be observed nevertheless. Only at high concentrations a totally homogeneous sedimentation of all particles takes place; this phenomenon is called ➔***zone sedimentation***.

Sweetland Filter

➔ ***Leaf filter***, patented in 1905 by E. Sweetland, with hanging filter leafs, arranged in a tank. Originally, the lower part of the vessel was unhinged for cleaning purposes; in present day designs the upper part of the vessel can be opened.

Swivel Beaker

Cup-shaped insert in discontinuously operating ➔***beaker centrifuges*** with a vertically rotating axis. At rest, when the beakers are charged, they hang vertically at the end of an arm, which is attached with pivoting joints to the rotational axis. During rotation they swivel sideways into a horizontal plane. In this manner, several beakers with ➔ ***suspension*** can be simultaneously processed.

Symmetric Membrane

➔ ***membrane*** with an uniform ➔ ***pores*** size across its thickness. Symmetrical membranes are mostly applied for ➔ ***microfiltration*** whereas ➔ ***asymmetrical membranes*** with an extremely fine-pored surface are applied for ➔ ***ultrafltration***.

T

t/V=f(V) Method

Experimental method to determine the → ***filter medium resistance*** and the → ***filter cake permeability*** in a → ***laboratory nutsche filter*** (→ ***Filtratest***) with the filter surface A. While a → ***filter cake*** is formed under constant pressure Δp, the suspension concentration (→ ***kappa-factor***) and the liquid viscosity η_L , the cumulative filtrate volume V is recorded over time t. Then the curve t/V-over-V is plotted.

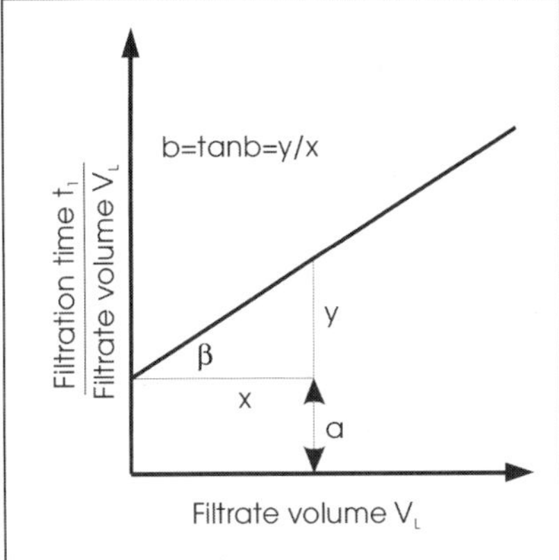

t/v over V for incompressible cakes

It shows for near incompressible cake formation a straight line and the filter medium resistance R_m can be calculated from the intercept a. The specific cake permeability p_c (or the specific cake resistance $r_c = 1/p_c$, respectively) are calculated from the slope b, as follows:

$$\frac{t}{V} = a + bV \qquad a = \frac{R_m \eta_L}{A \Delta p} \qquad b = \frac{\kappa \eta_L}{2 p_c A^2 \Delta p}$$

Generally, the specific cake resistance r_c has values between 10^{11} and $10^{16} m^{-2}$; at $10^{11} m^{-2}$ are extremely well filtering products (i.e. coarse salts, minerals) whereas $10^{16} m^{-2}$ characterizes very slowly filtering materials (bacteria, pigments).

Table Filter

Continuously working horizontal → ***vacuum filter***, which similar to a → ***disc filter*** is divided into a number of → ***filter cells*** with individual filtrate discharges. Although table filters are very similar to → ***bowl filters***, their rim edges are not rigidly connected with the filter disc, but instead is stationary with a sliding seal. The solids discharge straight horizontally through an opening in the wall via a → ***discharge screw***. Table filters are employed for easy to filter and rapidly settling → ***suspensions***.

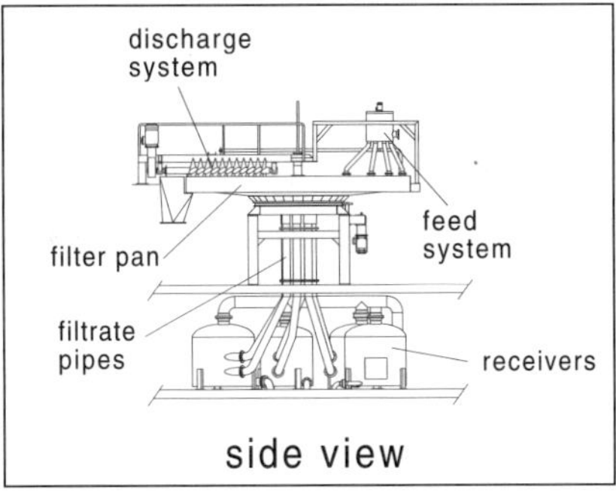

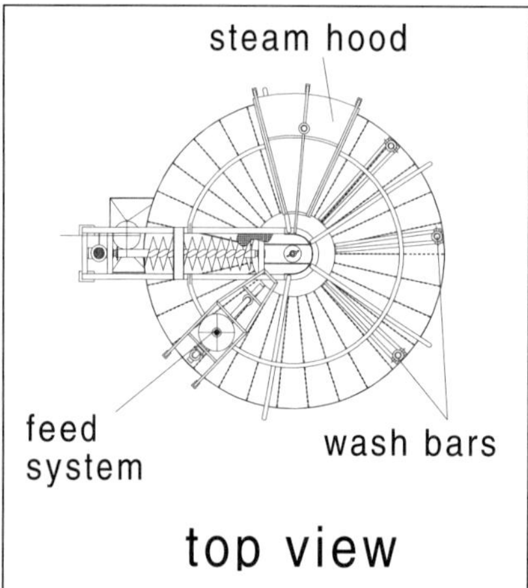

BOKELA ***high performance pan filter***

Tangled Fiber Fleece

➔ ***Fleece***

Taylor Vortex

Under certain operating conditions in a coaxial ➔ ***dynamic crossflow filter*** annular vortices develop, that rotate as pairs against each other. If an axial velocity component is superimposed analogous to continuous ➔ ***filtration***, then two continuously parallel and counter-rotating screw vortices evolve out of the individuals. This leads to an extremely complex velocity distribution.

Tension Roller

Deflection roller for the filter cloth in ➔ ***belt filters***, ➔ ***press filter automat*** or ➔ ***double belt presses***, with an adjustment for tightening of the circulating ➔ ***filter medium***.

Thermal Convection

Flows in a liquid caused by density differences in turn due to temperature gradients. Next to ➔ ***Brownian motion***, they are the main reason why individual particles with a diameter of less than approximately 1 µm cannot settle out in the earth's gravitational field.

Thermal Fixation

pre-treatment of synthetic fiber weaves, generally with hot air, to condition the cloth for dimensional stability during operation.

Thermal Loading Control Regulator

➔ ***Loading Regulator***

Thickener

➔ ***Gravity thickener***

Thickening filter

works either discontinuously or continuously to produce a highly concentrated, still free-flowing ➔ ***sludge*** instead of discharging only a demoistured ➔ ***filter cake***. They can be operated on the basis of ➔ ***cake filtration***, ➔ ***crossflow filtration***, or ➔ ***deep bed filtration***. Generally, the thickened sludge has to be further demoistured in a machine especially designed for that purpose. The thickening filter represents an alternative to gravitational sedimentation or ➔ ***centrifugal sedimentation***.

Thickening

denotes increasing the ➔ ***solids concentration*** of a diluted suspension to produce a still free-flowing ➔ ***sludge***. The thickening represents a typical pretreatment process in solid-liquid separation, most of the time followed by a further ➔ ***demoisturing*** of the sludge with suitable equipment. Thereby the generally high-grade equipment downstream is no longer burdened with large volumes or streams of liquid. Both settling and filtering processes can be utilized for the thickening, i.e. gravitational sedimentation, ➔ ***centrifugal sedimentation*** as well as ➔ ***cake filtration***, ➔ ***crossflow filtration***, and the ➔ ***deep bed filtration*** with regenerating beds.

Thin Layer Filtration

➔ ***Sieve Filtration***

Thixotrope

Non-➔ ***Newtonian flow*** behavior. With increasing duration of the acting shear forces the ➔ ***viscosity*** of the liquid decreases.

Three-Column Centrifuge

Discontinuously working ➔ ***centrifuge*** with perforated filter or solid bowl, a vertical rotational axis, and mounted on three spring-dampened legs for vibration absorption. The centrifuge drum itself is rigidly mounted in the spring-suspended housing. A typical example for a three-column centrifuge is a ➔ ***peeler centrifuge***. Three column centrifuges operate in the range of low 100 up to several 1,000 rpm and are furnished with approx. 0.5 up to 2m drum diameter. They are well suited for smaller batches of frequently changing products, as often required in the chemical industry, because they are accessible and easy to clean.

Three Phase Border

Linear contact formed by three different phases with a physical interface between each other, e.g. solid/liquid/gas, solid/liquid/liquid or liquid/liquid/gas.

Three Phase Decanter

➔ ***Decanter***, i.e. a ➔ ***solid bowl centrifuge***, which aside from solid separation can also separate two immiscible liquids. The liquids separate from each other due to their density difference into layers that are coaxially stacked around the rotational axis of the decanter, and are removed out of the process chamber by dedicated discharge devices such as ➔ ***peeling discs***. The applications for such machines are found, for example, in oil-saturated waste- water, or in the processing of vegetable oil.

Throughput

Quantity based on mass or volume which is able to pass through a separation apparatus per time unit. The throughput can be related to the amount of ➔ ***suspension*** fed, or the filtrate or solids flows produced, respectively. Both moist and dry solids can be meant in the case of solids throughput. If a specific throughput is

stated, then this value refers in addition to the employed ➔ ***filter area***. For example the specific throughput of ➔ ***rotary filters*** is quoted in (kg/m^{2}h).

Tilting Pan Filter

Quasi-continuously working vacuum filter of an older design made with individual suction box arranged horizontally in a circle around a ➔ ***central control valve***. The suction boxes are further turned stepwise according to a time-controlled program. Following the last demoisturing step they are tilted and the cake drops out of the filter cell.

Titer Reduction

➔ ***Bacteria Retaining Test***

Titus Pneumatics

➔ ***Peeling Pneumatics***

Titus Filter-Dryer

➔ ***Filter Reactor***

Torque

is a term used in mechanics defined as the product of force and lever arm which is the perpendicular distance from the axis to the line of action of the force acting on a revolving rigid object.

Total Separation Degree

Θ measures the solids m_G, separated out of a ➔ ***suspension***, in relation to the solids amount in the feed m_A:

$$\Theta = \frac{m_G}{m_A}$$

Transmembrane

indicates the direction perpendicular to the surface of a ➔ ***membrane*** from the ➔ ***concentrate*** to the ➔ ***permeate side*** of the medium.

Transtubular

indicates the direction parallel to the surface of a ➔ ***membrane*** of a cylindrical filter module from the point of entry of the ➔ ***suspension*** to the exit of the ➔ ***concentrate***.

Tray Belt Filter

is a special design of the ➔ ***belt filter*** by the DORR OLIVER company that features vacuum trays covered with a continuous ➔ ***filter medium***. This design is more elaborate than a basic ➔ ***vacuum*** filter with a ➔ ***rubber conveyor***, but it offers considerably more freedom in the selection of suitable filter media, and it operates like a continuous ➔ ***vacuum*** filter.

Tricanter

A ➔ ***decanter centrifuge*** for separating of a three-phase suspension, i.e. a solid and two immiscible liquids of different density, by the FLOTTWEG company. The separated liquids are drained off with two different discharge systems. There is an option to discharge one of the components with and the other without pressure. The position of the separation line between the liquids can be

adjusted with a height-variable ➔ ***peeling disc.***

Tromp Curve

➔ ***Fractional Grade Efficiency***

Tubular Centrifuges

Cylindrical sedimentation centrifuge of distinct slenderness ratio. They are operated with over flow until the solids collection space is filled up with separated solids. Tubular Centrifuges are used for separating extremely fine particles out of highly diluted suspensions with ➔ ***C-values*** of several 10,000's.

Tumbler Centrifuge

Continuously working ➔ ***selftransporting centrifuge*** with a conically opening and a cantilevered sieve basket. The transport of solids through this ➔ ***sliding centrifuge*** is accomplished with a tumbling motion of the sieve basket around its rotating axis. The machine is designed in such a way that at normal rotations the solids remain in place by adhesive friction. The solids are forced to slide intermittently by the tumbling movement to manage the otherwise difficult to control solids transport in sliding centrifuges.

Turbid Filtration Phase

occurs at the very first moment of filter cake formation. It is caused by fine solids particles that are able to pass through the ➔ ***pores*** of the ➔ ***filter medium*** before a ➔ ***bridge layer*** is formed, which serves as a filter agent for the remaining suspension.

Turbid Substances

➔ ***Colloids***

Turbidity

➔ ***turbid filtration phase***

Turbodrain

Continuously working gravitational belt thickener by the BELLMER company for upgrading flocculated thin slurries. Plowshare-like baffles are moving across a horizontal demoisturing tray in order to reposition the sludge layer over and over again and to tear open new flow channels for the draining liquid. The machine serves for instance as a pre-treater of ➔ ***suspensions*** for the subsequent demoisturing on a ➔ ***double belt press.***

Turbulent Flow

sets in when the critical ➔ ***Reynold's number*** is exceeded as minor flow fluctuations are no longer attenuated and random turbulences disturb the fluid movement. These are small liquid or gas eddies that move volume elements diagonally to the flow direction, so that the liquid or gas layers are being mixed.

Twill Lace

Special form of ➔ ***twill weave,*** at which warp and weft (➔ ***weave***) have different diameters. This makes the ➔ ***weave*** stronger; in addition the ➔ ***mesh width*** can be modified.

Twill Weave

Special form of thread weave in a ➔ ***weave***. A weft thread is bound over respectively two or more warp threads or reverted (weft- respectively warp thread). The interlacing points rise in a twill weave in uninterrupted diagonals, forming the twill marks. Within the ➔ ***pattern repeat*** one or more weave marks can exist. In a satin weave the interlacing points do not contact each other, so that a smooth, surface without structure is formed. Twill weaves often have uneven sides, if warp and weft are not equally distributed on the upper- and underside of the cloth.

Two-Phase Flow

in solid-liquid separation technology means often a mixed flow of liquid and gas in a ➔ ***bulk***. In contrast to tubular two-phase flows, these fluids move to a large extent independently from each other in the fine pore channels of particle ➔ ***bulks***. Thus the liquid volume is assigned to the solids when the gas flow is evaluated since the shear stress induced by the ➔ ***friction*** of the gas into the liquid is generally smaller by several orders of magnitude than the ➔ ***capillary pressure*** which retains the liquid.

Ultrafiltration

A membrane-type filtration, often executed as ➔ ***crossflow filtration*** for separating submicron-sized particles or dissolved macromolecules. The ➔ ***cut point*** of ultrafiltration membranes is often characterized by the so-called ➔ ***MWCO*** (***M***olecular ***W***eight ***C***ut ***O***ff).

Underpressure

Lowered absolute pressure acting against a surrounding atmospheric pressure. This underpressure, also known as vacuum, can be decreased in low vacuum filtrations only down to the vapor pressure of the liquid, i.e. the underpressure for water at room temperature and sea level can be decreased down to 0.2bar a.

Unfiltrate

If the term ➔ ***suspension*** does not make sense anymore for a liquid to be separated such as in a ➔ ***sterile filtration***, it may be called an unfiltrate.

Upgrading

➔ ***Thickening*** of ➔ ***suspensions***, meaning an increase of the ➔ ***solids concentration***.

Vacuum

➔ *Underpressure*

Vacuum Belt Filter with Rubber Conveyor Belt

Special design form of a ➔ ***belt filter***. A vacuum belt filter with rubber conveyor belt works fully continuously. The ➔ ***filter medium*** is carried by a continuous rubber conveyor belt underneath serving for the transport function as well as for mechanical stability. The belt has filtrate run-off grooves diagonally to the centerline and drilled wholes for the drainage of ➔ ***filtrate*** by suction. Vacuum belt filters with rubber conveyor belts cannot be employed in all fields without limitations, because frequently ➔ ***suspensions*** are not compatible with rubber, especially in the chemical industry. For this reason other belt filters have been designed, such as the ➔ ***belt filter*** with reversing vacuum trays or the continuously working ➔ ***tray belt filter***.

Vacuum Filter

Discontinuously or continuously working class of cake filters that drain filtrate with underpressure applied in the ➔ ***filter cell*** on the underside side of the ➔ ***filter medium***.

Vallez-Filter

A ➔ ***leaf filter*** patented by E. Vallez in 1916 consisting of a horizontal vessel with round filter leaves mounted vertically on a common hollow shaft. The shaft with the filter leafs rotates slowly during the cleaning and the cake is removed with spray nozzles, i.e. in wet manner. The Vallez-filter was the first pressure leaf filter suitable to be cleaned in closed condition.

Van-der-Waals Forces

are bonding forces between atoms and molecules that are not based on a complete or even partial electron transition. They result from interactions between fluctuating electrical dipoles. During the ➔ ***coagulation*** of particles Van-der-Waals forces can come into play only if the electrostatic repulsion between the particles is small enough.

VC-Filter

Horizontally arranged tube filter press with press membrane.

Vertical Pendulum Filter Centrifuge

Discontinuously working ➔ ***filter centrifuge***, whose perforated drum can oscillate freely on an top-mounted, vertically rotating shaft in a solid housing. Self-centering is forced by rotation. The rotating shaft is mounted on an appropriate frame or bridge frame. Vertical pendulum filter centrifuges are

conceived for bottom discharge and mainly used in the sugar industry for separating and ➔ ***washing*** crystalline sugar. The cake is discharged with peeler knives or a sharp edge of the drum where the solids cake breaks off and falls out of the drum bottom.

Vibrating Screen

Rectangular sieve, which is oscillated by a vibrating motor. The moist feed material is charged onto the narrow front side and simultaneously transported and demoistured by the sieve movement. Vibrating screens are typically employed for a large-scale sand demoisturings. The processing can be influenced by changing the sieve slope, and the frequency and amplitude of the vibration.

Vibration Absorption Plate

Massive steel or concrete plate, which is attached for instance under horizontal ➔ ***peeler centrifuges*** in order to increase their total weight for vibration dampening. The complete aggregate of ➔ ***centrifuge*** and vibrating absorption plate is mounted again on spring elements which normally neutralize the oscillations transmitted into the foundation.

Vibration Damper

Spring elements often attached under ➔ ***centrifuges*** to shield the foundation from the vibrations generated by the operation.

Vibrational Density

➔ ***Bulk Density***

Vibratory Centrifuge

Continuously working ➔ ***sieve centrifuge*** with a cantilevered filter drum, which is enlarged conically to the outside. The ➔ ***suspension*** is added at the bottom center of the cone and is conveyed by coaxially superimposed oscillations to the outer edge of the cone, where the demoistured solids are discharged. Vibratory centrifuges have a similar field of application as ➔ ***vibrating screens***; they are well suited for easy to demoisture materials at high ➔ ***suspension concentration*** and large flow rates. They are built with drum diameters of approx. 0.5÷1.5m, operated with ➔ ***C-values*** in the region of 60÷150, and have a throughput of up to 350m.t./hr.

Viscosity

Viscousness of a fluid expressed as either ➔ ***dynamic viscosity*** or ➔ ***kinematic viscosity***.

Viscousness

➔ ***Viscosity***

Void Ratio

The void ratio e is defined as the void volume V_V divided by the solids volume V_s of a ➔ ***bulk***.

$$e = \frac{V_V}{V_s}$$

The void ratio is related to ➔ ***porosity***.

Volume Concentration

➔ *Solids Content*

Vortex Finder

A pipe, immersed from above into the center of a ➔ ***hydrocyclone***, through which the main portion of the suspension liquid is discharged. All particles smaller than the ➔ ***cut-size*** are also carried by the ➔ ***secondary vortex*** out of the cyclone.

Vortex Flow

➔ *Potential Vortex*

W.A.P.

→ ***Wring Alternating Press***

Warp

→ ***Weave***

Wash Degree

Residual portion of a substance dissolved in the → ***mother liquor*** that is to be washed out of a → ***filter cake***. The wash degree is normalized by assigned values of 1 before the washing and 0 after the total removal of the undesired substances, respectively.

Washing

→ ***Cake Washing***

Washing Liquid

Used for the removal of undesired substances, which are dissolved in the → ***mother liquor*** of a → ***bulk***. The → ***bulk*** can either be permeated by the washing liquid (see → ***displacement washing***) or the demoistured → ***bulk*** is resuspended in it separately and filtered again. The latter case causes a dilution effect (see → ***dilution washing***).

Washing Out

→ ***Washing***

Washing Pipe

Horizontal pipe with a single nozzle at the end to spray a → ***washing liquid*** on a → ***filter cake*** in → ***pusher centrifuges*** *etc.*

Washing Spray

Device especially applied at → ***drum filters*** for the → ***washing*** of a → ***filter cake***. Pipes with spray nozzles running across the drum distribute the → ***washing liquid*** as uniform as possible film over the cake surface.

Washing Ratio

Characteristic measure in → ***displacement washing*** for the consumption of washing liquid. The washing ratio relates the volume of washing liquid to the pore volume in a → ***filter cake***. Ideally, the ratio would be unity as the pore volume would be replaced just once by washing liquid. In reality, however, washing ratios can be considerably higher.

Water Content

→ ***Residual Moisture***

Water Value

Reference value for the permeation rating of → ***filter media***. It is determined with particle free water under defined conditions and by relating the passed

volume to the surface. The water value serves mainly for the comparative ranking of filter media; it is of little value regarding the filtration performance with an actual ➔ ***suspension***.

Wear

➔ *Abrasion*

Weave

Textile material, fabricated in a weaving mill, from two thread systems crossing each other in a rectangular pattern in a weave. Transversely to the in longitudinal direction running warp (warp threads) the weft is woven in and out repeatedly at the weave's edges or ledges.

Wedge Wire Filter

Discontinuously working, candle-shaped filter for purifying large liquid volumes with low particle contamination. The filtration process on the slit-shaped filter openings is stopped when a pre-set value for the developing pressure loss is reached. Then the wedge filter is regenerated mechanically by means of brushes or scrapers, mostly while filtering. The gap width of the filter sieve has to be adjusted for the particle size to be retained.

Wedge Wire Sieve

Metallic, rigid and non-woven ➔ ***filter medium*** with slot-shaped openings. Wedge wire sieves are used as filter medium for instance in screen centrifuges or ➔ ***worm extruders***. These sieves are considerably more robust and resistant against abrasion than a weave.

Wedge Zone

A second demoisturing section in a ➔ ***double belt press*** following the pre-demoisturing ➔ ***straining zone*** where the ➔ ***bulk*** is compacted by additional pressing and shearing on its way through the roller press system.

Weft

➔ *Weave*

Weir Disc

Device used in sedimentation centrifuges (see ➔ ***decanter***) for the liquid level control in the machine. The weir disc can be designed either for a fixed level or through height-adjustable radial slots for a variable filling.

Wetting Angle

➔ *Wetting* ➔ *Contact Angle*

Wetting

Spreading of a liquid on a solid surface. A liquid that spreads spontaneously across the entire body is called a completely wetting or ➔ ***spreading liquid***. The ➔ ***contact angle*** at the ➔ ***three-phase boundary*** of solid, liquid and gas is in this case 0°. Glass, water and air form a completely wetting system. If the contact angle is smaller than 90° a liquid shows wetting behavior. Contact angles larger than 90° define the non-wetting class; the system glass-mercury-air would be an example.

Wetting is of importance in solid-liquid separation due to its direct influence on the capillary pressure and therefore it plays a role in ➔ ***de-moisturing*** of ➔ ***filter cakes***.

Wing-Shear Strength

Test method for measuring the aggregate strength of moist ➔ ***bulk materials*** designated for landfilling. A wing probe is inserted into a moist ➔ ***bulk material*** sample that has been prepared according to specifications and the torque needed to rotate the probe is measured. A minimum value of approx. $2N/m^2$ suffices for landfilling. The wing-shear strength depends on the type of material and its ➔ ***residual moisture***.

Worm Extruder

Continuously working ➔ ***press filter apparatus*** made up of a screw in a perforated cylinder with contracting cross sectional area. The screw conveys the feed material axially and compresses it continuously. The separated liquid permeates the ➔ ***filter medium*** (e.g. gap sieve) and flows to the outside. Worm extruders are preferably employed with substances that are coarse, compressible and often fibrous in addition. Known applications extend as far as to the separation of liquid manure.

Worm Screen Centrifuge

Continuously working ➔ ***filter centrifuge*** with a conically widening, cantilevered sieve basket, in which the solids transport is facilitated with a conveyer screw rotating at a ➔ ***differential number of revolutions***. Worm screen centrifuges have drum diameters of approx. 0.2÷1.0m and operate with ➔ ***C-values*** of approx. 200÷3000. Their suspension throughput is approx. 1.0÷80m.t./hr. Particles to be separated should be larger 100µm. ➔ ***Cake washing*** is limited due to the drum internals.

Wound Module

➔ ***Spiral Module***

Wrapped Module

➔ ***Spiral Module***

Wring Alternating Press

Special design of a discontinuously working piston filter press developed by the BOKELA and Siempelkamp companies with pressing powers of up to approx. 100bar. A novelty of these presses are the parallel arranged drainage filter media that are inserted in the process chamber. Between the media the sludge is pressed to a thin layer of 1÷2mm thickness. During pressing the drainage cloths fold up and create in this manner a multitude of channels through which the released ➔ ***filtrate*** can readily drain off. The thin layer filtration applied in this press significantly accelerates the pressing process in comparison to ➔ ***filter presses***. In addition, the extremely high pressing power produces considerably lower equilibrium moisture. When the press is opened the drainage cloths are stretched again and the pressed cake drops out in pieces. Residual cake solids still adhering can be shaken off by a twisted swaying of the cloth package.

Presently, machines are built with a total filter area of up to $58 m^2$.

Drainage module of the ***wring press***

Wring Press

➔ ***Wring Alternating Press***

Wyckoff & Botset Equation

The permeability of ➔ ***filter cakes*** changes for both the liquid and the gaseous-phase flows with a change in the ➔ ***saturation degree*** during a gas differential pressure demoisturing of the system. As for the permeability of the gaseous phase $p_g(S)$ as a function of the saturation degree S, it is according to Wyckoff&Botset the product of the single phase permeability p_c and a relative gas permeability $p_{rel,g}(S)$ ranging from 0 to 100%:

$$p_g(S) = p_c p_{rel,g}(S)$$

The permeability for the liquid phase $p_L(S)$, in turn, also as a function of the saturation degree S, is formed as the product of the single phased permeability p_c and a relative liquid permeability $p_{rel,L}(S)$ which also can attain values between 0 and 100%:

$$p_L(S) = p_c p_{rel,L}(S)$$

The function $p_{rel,L}(S)$ can be described by the ➔ ***Brutsaert equation***.

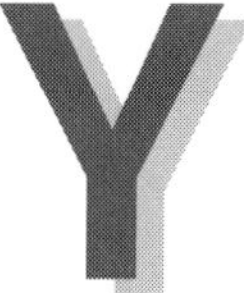

Yield Point

The yield point of a body under strain is defined as the tension at which plastic deformation begins with a measurable velocity.

Z

Zeta Meter

Instrument for the measurement of the ➔ ***zeta potential*** of particles suspended in an electrolytic fluid.

Zeta Potential

The zeta potential is the gradient of the electro-kinetic potential at a solid-liquid phase border. The surface of a suspended particle is generally electrically charged (often negative). An ionic layer that surrounds this surface is formed in the suspension liquid from dissociated molecules with opposite charges. This layer may compensate to a large degree but not completely the surface charge of the particles. Complete neutralization occurs with an additional diffuse liquid layer around the particle, which still displays a slight surplus of the respective counter ions. Charge equilibrium exists only outside this ➔ ***double layer***. Zeta potentials can be measured by a forced tangential displacement of the mobile double layer, with values ranging typically of +40÷ -50mV.

Zone Sedimentation

➔ ***Agglomeration*** or ➔ ***flocculation*** of the solid particles can intensify the hindered sedimentation (see ➔ ***Swarm Sedimentation***). In the region of low solids concentrations an excess sedimentation velocity can be observed caused by the formation of hydrodynamic complexes. At higher concentrations the sedimentation velocity decreases strongly as a swarm is formed, in which the velocity determining particle properties like specific density, size and shape lose their influences. In this region, the resulting hindered sedimentation can be described with the ➔ ***Richardson & Zaki equation***. The hindered sedimentation is beneficial in sedimentation equipment because it forms a sharp ➔ ***sedimentation front***.

We about us

We perform tailor-made solutions

BOKELA

For more than 15 years we are playing an important role in the process industry for solid/liquid separation.

Competence, creativity, reliability and flexibility – these features are characterising us. Our high performance technologies are appreciated by customers world wide. With pioneering innovations we received greatly valued innovation awards.

We supply tailor-made, complete plants and equipment for solid/liquid separation. Revamping of running separation plants and our comprehensive service made us well-known.

The present BOKELA SLS LEXICON may demonstrate our expertise.

BOKELA

Ingenieurgesellschaft für
Mechanische Verfahrenstechnik mbH

Gottesauer Str. 28
76131 Karlsruhe
Phone 07 21/9 64 56-0
Fax 07 21/9 64 56-10
E-Mail bokela@bokela.com
www.bokela.com